Cambridge Tracts in Mathematics and Mathematical Physics

No. 9

Invariants of Quadratic Differential Forms

INVARIANTS
OF
QUADRATIC DIFFERENTIAL FORMS

by

J. EDMUND WRIGHT, M.A.
Fellow of Trinity College, Cambridge
Associate Professor of Mathematics, Bryn Mawr College, U.S.A.

Cambridge:
at the University Press
1908

CAMBRIDGE
UNIVERSITY PRESS

University Printing House, Cambridge CB2 8BS, United Kingdom

Cambridge University Press is part of the University of Cambridge.

It furthers the University's mission by disseminating knowledge in the pursuit of education, learning and research at the highest international levels of excellence.

www.cambridge.org
Information on this title: www.cambridge.org/9781107493933

First published 1908
Re-issued 2015

A catalogue record for this publication is available from the British Library

ISBN 978-1-107-49393-3 Paperback

PREFACE

THE aim of this tract is to give, as far as is possible in so short a book, an account of the invariant theory connected with a single quadratic differential form. It is intended to give a bird's eye view of the field to those as yet unacquainted with the subject, and consequently I have endeavoured to keep it free from all analysis not absolutely necessary.

It will be found that the rest of the tract is independent of Chapters III and IV. These chapters are included so as to give an account, as far as possible complete, of the various methods that have been applied to the subject. The most successful method is that outlined in the remainder of the book. This method, begun by Christoffel, owes its modern development mainly to Ricci and Levi-Civita, and it is hoped that this tract may induce some of its readers to turn to their papers.

J. EDMUND WRIGHT.

TRINITY COLLEGE.
July 1908.

CONTENTS

INVARIANTS OF QUADRATIC DIFFERENTIAL FORMS

INTRODUCTION

1. In order to discuss in detail the geometry of a plane it is convenient to introduce coordinates. A point in the plane has two degrees of freedom, and therefore to determine it two independent conditions must be satisfied. These conditions may be that two independent quantities* (*e.g.* the distances from two fixed points) take the values u, v at the point, and we then say that the coordinates of that point are u, v. If we suppose one coordinate given, the locus of the point will be a certain curve, *e.g.* $u = \text{const.}$, and, generally, a curve in the plane is given by a functional relation $\phi(u, v) = 0$.

For the metrical geometry of the plane we need an expression for the distance between any two points in terms of the coordinates of the points. Theoretically this can be calculated if the distance between an arbitrary point (u, v) and any neighbouring point $(u + du, v + dv)$ is known. Suppose that (x, y) and $(x + dx, y + dy)$ are rectangular cartesian coordinates of the two points, then x, y are functions of u, v; also if ds denote the distance between the two points,

$$ds^2 = dx^2 + dy^2,$$

or

$$ds^2 = Edu^2 + 2Fdudv + Gdv^2,$$

where

$$E = \left(\frac{\partial x}{\partial u}\right)^2 + \left(\frac{\partial y}{\partial u}\right)^2, \quad F = \frac{\partial x}{\partial u}\frac{\partial x}{\partial v} + \frac{\partial y}{\partial u}\frac{\partial y}{\partial v}, \quad G = \left(\frac{\partial x}{\partial v}\right)^2 + \left(\frac{\partial y}{\partial v}\right)^2.$$

We have in fact a quadratic form in the variables du, dv, for ds^2, and the coefficients of this form are functions of u, v. If E, F, and G are

* These quantities have not necessarily any obvious geometrical significance.

given it is possible to determine the equations of the straight lines of the plane in terms of u, v, to find the angle between any two of its curves, and, generally, to develope its metrical geometry. Now if the system of coordinates is given, E, F, G can be determined, and the converse question at once arises, namely, if three functions E, F, G are given as three arbitrary functions of u, v, is it possible to take u, v as coordinates of the points in the plane so that the element of length shall be given by

$$ds^2 = Edu^2 + 2Fdudv + Gdv^2\,?$$

It appears that this is not possible unless a certain relation

$$K \equiv \frac{1}{2\sqrt{EG-F^2}}\left\{\frac{\partial}{\partial u}\left[\frac{F}{E\sqrt{EG-F^2}}\frac{\partial E}{\partial v} - \frac{1}{\sqrt{EG-F^2}}\frac{\partial G}{\partial u}\right]\right.$$
$$\left. + \frac{\partial}{\partial v}\left[\frac{2}{\sqrt{EG-F^2}}\frac{\partial F}{\partial u} - \frac{1}{\sqrt{EG-F^2}}\frac{\partial E}{\partial v} - \frac{F}{E\sqrt{EG-F^2}}\frac{\partial E}{\partial u}\right]\right\} = 0$$

is satisfied for all values of u, v. This condition is also sufficient.

If, instead of limiting ourselves to a plane, we consider any surface in three dimensional space, we have again two coordinates for any point; the element of length is given as before by

$$ds^2 = Edu^2 + 2Fdudv + Gdv^2,$$

and there is a surface corresponding to any arbitrary functions E, F, G of u, v. (The particular case $EG = F^2$ is excluded.) It appears however that for a given surface the expression K has the same value at any given point on it, whatever coordinates u, v are chosen. Let u, v and u', v' denote any two sets of point coordinates on the surface, then $u' = f(u, v)$, $v' = \phi(u, v)$, where f and ϕ are arbitrary functions of u, v, and we have the theorem:

If by any transformation

$$u' = f(u, v),\; v' = \phi(u, v),$$

$$Edu^2 + 2Fdudv + Gdv^2 \textit{ becomes } E'du'^2 + 2F'du'dv' + G'dv'^2,$$

then $K = K'$, where K' is K in the accented variables.

2. Definition of a differential invariant.

Any function of E, F, G and their derivatives satisfying this condition is called a *differential invariant* of the form

$$Edu^2 + 2Fdudv + Gdv^2.$$

The idea of a differential invariant may be extended by taking account of any families of curves on the surface, say $\phi(u, v) = \text{const.}$,

$\psi(u, v) = \text{const.}$, etc. When we transform to new variables u', v' we have

$$Edu^2 + 2Fdudv + Gdv^2 = E'du'^2 + 2F'du'dv' + G'dv'^2,$$
$$\phi(u, v) = \phi'(u', v'),$$
$$\psi(u, v) = \psi'(u', v'), \text{ etc.},$$

and a differential invariant is defined as a function of u, v, E, F, G, ϕ, ψ, and their derivatives (u, v being regarded as independent variables) that has the same value whether written in the original or in the transformed variables.

Invariants which involve only u, v, E, F, G and their derivatives are called *Gaussian* invariants, while those which involve also derivatives of ϕ, ψ, etc. are called differential parameters. Thus for example K is a Gaussian invariant and

$$\Delta\phi \equiv \frac{1}{EG - F^2}\left\{E\overline{\frac{\partial\phi}{\partial v}}\Big|^2 - 2F\frac{\partial\phi}{\partial u}\frac{\partial\phi}{\partial v} + G\overline{\frac{\partial\phi}{\partial v}}\Big|^2\right\}$$

is a differential parameter of the quadratic differential form

$$Edu^2 + 2Fdudv + Gdv^2.$$

If the quadratic form is interpreted as the square of the element of length of a surface in space, K and $\Delta\phi$ have also geometrical interpretations. K is the Gaussian or total curvature of the surface, and if $\Delta\phi = 1$, the curves $\phi = \text{const.}$ are the orthogonal trajectories of a family of geodesics on the surface.

The extension of these ideas from two to m variables is immediate, and the quadratic differential form in m variables may be regarded as the square of the element of length in the most general m dimensional manifold. The main point is that invariants are independent of the particular choice of coordinates, in other words they are intrinsically connected with the manifold itself. The course of ideas is as follows. We start with a given manifold, which possesses certain properties. Some of these may be independent of each other, some may be consequences of certain others, and there are relations connecting these. We may develope the discussion on the lines of pure geometry, but we are compelled, sooner or later, to appeal to algebraic methods. These methods involve the introduction of coordinates, and properties of the manifold are then expressed by means of algebraic equations. An algebraic expression has some interpretation in the manifold taken together with the coordinate frame used, and a complication has been introduced, for the discussion will now involve those additional properties which are not intrinsic to the manifold, but arise out of the particular coordinate frame chosen. If however we work only with invariants, we avoid this latter class of properties and are able at the same time to use the powerful methods of analysis. The geometry of the manifold thus breaks up into two parts:

(i) The determination of all invariants and all relations connecting them.

(ii) The geometrical interpretation of all these invariants in the manifold.

3. So far there have been considered only invariants arising through the quadratic form that is equal to ds^2. These are all the invariants when we consider the manifold in itself, but if we suppose it existing in, say, Euclidean space of higher dimensions we introduce other invariants connected with the relation of that space to the manifold. For example, in the case of a surface in space, the totality of invariants is only given when *two* quadratic forms are taken account of, the additional one being that which determines the normal curvature at any point of the surface. The surface, in fact, is not intrinsically determinate by means of the single form, but may be bent provided there is no tearing or stretching, or as we say, it may be *deformed*, without alteration to ds^2. (For instance any developable may be deformed into a plane.) The discussion when there are two or more forms is similar to that when there is only one. The invariants arising from the single form are called deformation invariants.

Thus far it is suggested that the invariants are essentially connected with differential geometry. This is by no means the case. They are connected with a certain form, and any interpretation of this form leads to a corresponding interpretation for the invariants.

Consider in fact any dynamical configuration with Lagrange coordinates $u_1, u_2, \ldots, u_n$. The kinetic energy of this system is $\sum_{r,s=1}^{n} a_{rs}\dot{u}_r\dot{u}_s$, where a_{rs} is a function of the variables u, and dots denote derivatives with regard to the time. By a new choice of coordinates we effect a transformation of exactly the same type as that already considered, and again we have a series of invariants of a quadratic form, and these are those quantities which are dependent on the configuration itself as distinct from the particular system of coordinates.

CHAPTER I

HISTORICAL

4. Group.

Invariance necessarily carries with it the idea of a transformation. Suppose we have a set of transformations in any variables whatever, and suppose that each of the set leaves a certain function of these variables invariant, then any transformation compounded of two or more of the set will also leave that function invariant. If any such transformation as this is not one of the original set we add it to that set, and we may thus continue adding new transformations until we reach a closed set, that is one such that if you apply in turn any two of its transformations the result is another of its transformations. Such a set is called a GROUP, and it is clear that any invariant whatever is invariant under a *group* of transformations.

5. In the case considered in the preceding pages there are a certain number of quadratic differential forms

$$\sum_{r,s=1}^{n} a_{rs} dx_r dx_s,$$

together with a certain number of functions $\phi(x_1, \ldots, x_n)$, and the group of transformations $x_i = x_i(y_1, \ldots, y_n)$, $(i = 1, \ldots, n)$, and we suppose that under a member of this group $\sum_{r,s=1}^{n} a_{rs} dx_r dx_s$ becomes $\sum_{r,s=1}^{n} a'_{rs} dy_r dy_s$, and that ϕ becomes ϕ'. Then there are deducible relations for a'_{rs}, ϕ', and their various derivatives with respect to the y's, and for $dx_1, \ldots, dx_n$ in terms of the original magnitudes a_{rs}, ϕ, etc. In other words there exists a set of transformations for all the variables mentioned. It may be proved that this set is a group, and this group is said to be *extended* from the original group. Our problem is the determination of all the invariants of this extended group.

6. Christoffel.

There have been three main methods of attack. The first, historically, is by comparison of the original and transformed forms, and in this way invariants are obtained by direct processes. The fundamental work in this direction is due to Christoffel* (1869), though the first example of an invariant, the quantity K, was given by Gauss† in 1827. Invariants which involve the derivatives of the functions are called *differential parameters.* Lamé‡, using the linear element in space given by $ds^2 = dx^2 + dy^2 + dz^2$, gave this name to the two invariants

$$(\Delta_1\phi)^2 \equiv \left(\frac{\partial\phi}{\partial x}\right)^2 + \left(\frac{\partial\phi}{\partial y}\right)^2 + \left(\frac{\partial\phi}{\partial z}\right)^2$$

$$\Delta_2\phi \equiv \frac{\partial^2\phi}{\partial x^2} + \frac{\partial^2\phi}{\partial y^2} + \frac{\partial^2\phi}{\partial z^2},$$

and Beltrami § adopted it for the invariants that he discovered, those involving first and second derivatives of a function ϕ, taken with a form in two variables.

In the course of Christoffel's work there arise certain functions $(ikrs)$; these were originally found by Riemann in 1861 in his investigations on the curvature of hypersurfaces. For a surface in space they reduce to the one quantity K.

7. Ricci and Levi-Civita.

To Christoffel is due a method whereby from invariants involving derivatives of the fundamental form and of the functions ϕ may be derived invariants involving higher derivatives. This process has been called by Ricci and Levi-Civita *covariant derivation*, and they have made it the base of their researches in this subject. These researches have been collected and given by them in complete form in the *Mathematische Annalen* ||, and on their work they have based a calculus which they call *Absolute differential calculus.* They give a complete solution of the problem, and show that in order to determine all differential invariants of order μ, it is sufficient to determine the algebraic invariants of the system:

(1) The fundamental differential quantic,

* *Crelle*, Vol. 70 (1869) p. 46.

† *Disquisitiones generales circa superficies curvas.*

‡ *Leçons sur les coordonnées curvilignes* (1859).

§ Darboux, *Théorie générale des surfaces*, Vol. III. pp. 193 *sqq.* gives an account of Beltrami's work together with a bibliography.

|| *Math. Ann.* Vol. 54 (1901) pp. 125 *sqq.*

(2) The covariant derivatives of the arbitrary functions ϕ up to the order μ,

(3) A certain quadrilinear form G_4 and its covariant derivatives up to the order $\mu - 2$*.

8. Lie.

The second method is founded on the theory of groups of Lie, and is a direct application of the theory given in his paper *Ueber Differentialinvarianten* †. This theory involves the use of infinitesimal transformations, and the invariants are obtained as solutions of a complete system of linear partial differential equations. Our problem is discussed shortly by Lie ‡ himself, for the case $n=2$. Żorawski § considered this case in detail, and gave the invariants of orders one and two. He also treated the question of the number of functionally independent invariants of any order. C. N. Haskins ‖ has determined the number of functionally independent invariants of any order. Forsyth ¶ has obtained the invariants of orders one two and three for a quadratic form in three variables, and of genus zero, that is to say, for ordinary Euclidean space. He has also obtained the invariants of the first three orders for any surface in space **, that is, for two quadratic forms in two variables, one perfectly general, and the other connected with it by certain differential relations. The problem for the differential parameters has been solved by this method by J. E. Wright ††.

9. Maschke.

The third method is due to Maschke ‡‡, who has introduced a symbolism similar to that for algebraic invariants. He developes processes similar to that of transvection, whereby an endless series of invariants may be constructed.

* *loc. cit.* p. 162.

† *Math. Ann.* Vol. 24 (1884) pp. 537 *sqq.*

‡ *loc. cit.*

§ *Ueber Biegungsinvarianten* in *Acta Math.* Vol. XVI (1892–1893) pp. 1–64.

‖ *Trans. Amer. Math. Soc.* Vol. III (1902) pp. 71, 91; also *ib.* Vol. V. (1904) pp. 167, 192.

¶ *Phil. Trans.* Series A, Vol. 202 (1903) pp. 277–333.

** *Phil. Trans.* Series A, Vol. 201 (1903) pp. 329–402.

†† *Amer. Journ. of Math.* Vol. XXVII (1905) pp. 323–342.

‡‡ *Trans. Amer. Math. Soc.* Vol. I (1900) pp. 197, 204; and Vol. IV (1903) pp. 445–469.

The geometrical interpretation of the invariants has been discussed at length by Forsyth*, and a considerable part of the work of Ricci and Levi-Civita deals with geometrical applications.

A general account of the whole subject was given by Maschke† at the St Louis Exposition, 1904.

An account will now be given of these three methods.

* See his two papers already quoted, and *Rendiconti del Circolo Matematico di Palermo*, Vol. 21 (1906) pp. 115–125.

† *St Louis Congress of Arts and Sciences*, Vol. I. pp. 519, 530.

CHAPTER II

THE METHOD OF CHRISTOFFEL

10. The quadratic form in two variables.

Let there be two quadratic forms $F \equiv adx^2 + 2bdxdy + cdy^2$ and $F' \equiv AdX^2 + 2BdXdY + CdY^2$, and suppose that x, y may be expressed as functions of X, Y so that when these values are substituted in F it becomes F'. We have then

$$adx^2 + 2bdxdy + cdy^2 = AdX^2 + 2BdXdY + CdY^2.$$

In this we write $dx = \frac{\partial x}{\partial X} dX + \frac{\partial x}{\partial Y} dY$, with a similar expression for dy, and the equation takes the form

$$PdX^2 + 2QdXdY + RdY^2 = 0$$

where P, Q, R are certain functions of x, y, X, Y and the derivatives of x, y with regard to X and Y. Now X, Y are independent variables and therefore there exists no relation among the differentials dX, dY, and hence P, Q, R are all zero. Thus the necessary and sufficient conditions in order that F shall be transformable into F' are $P=0$, $Q=0$, $R=0$, or written at length.

$$a\left(\frac{\partial x}{\partial X}\right)^2 + 2b\frac{\partial x}{\partial X}\frac{\partial y}{\partial X} + c\left(\frac{\partial y}{\partial X}\right)^2 = A,$$

$$a\frac{\partial x}{\partial X}\frac{\partial x}{\partial Y} + b\left(\frac{\partial x}{\partial X}\frac{\partial y}{\partial Y} + \frac{\partial x}{\partial Y}\frac{\partial y}{\partial X}\right) + c\frac{\partial y}{\partial X}\frac{\partial y}{\partial Y} = B,$$

$$a\left(\frac{\partial x}{\partial Y}\right)^2 + 2b\frac{\partial x}{\partial Y}\frac{\partial y}{\partial Y} + c\left(\frac{\partial y}{\partial Y}\right)^2 = C.$$

These three are differential equations of the first order for x, y as functions of X, Y. If they can be solved their solution gives the transformation whereby F is changed into F'. Now there are three equations, and they involve only two dependent variables x, y; hence they cannot in general co-exist unless there be relations between

a, b, c, A, B, C and their derivatives. Our first problem is to find the conditions in order that they may co-exist.

By differentiation we obtain six equations in the six second derivatives of x, y, and these may be solved for the second derivatives in question. If the original three are differentiated twice there are obtained nine equations involving third derivatives, and by means of the equations for second derivatives these may be reduced to a form in which they involve first and third derivatives only. There are only eight third derivatives of two functions x, y, each of two variables X, Y, and therefore by eliminating them from this last set of equations we get a new equation, which, since it involves first derivatives only, must be added to the original three equations. It happens that from these four equations the first derivatives can be eliminated and thus there is given a relation between a, b, c, A, B, C and their first and second derivatives. This relation is precisely $K = K'$.

We can now proceed step by step to find the equations involving higher derivatives of x, y, and then by elimination to find other relations among the coefficients a, b, c, A, B, C and their derivatives. In the case considered, that of two independent variables, these relations all follow from the equivalence of K and K'.

11. The quadratic form in n variables.

The general quadratic in n variables may be treated in exactly the same manner; the statement of the work is much simplified by the use of certain abbreviations which we proceed to define.

The form F itself is written $\sum_{r,s} a_{rs}dx_r dx_s$ and the form F' is $\sum_{r,s} a'_{rs}dy_r dy_s$, the summation being always from 1 to n for each of the letters under the sign of summation. The y's are taken as the independent variables, and the x's are assumed to be functions of these. The determinant of n rows and columns, whose elements are the quantities a_{rs}, is called a. The cofactor of the rth row and sth column in a is written Δ_{rs}. The quantity

$$\tfrac{1}{2}\left[\frac{\partial}{\partial x_h} a_{gk} + \frac{\partial}{\partial x_g} a_{hk} - \frac{\partial}{\partial x_k} a_{gh}\right] \text{ is written } [gh, k],$$

and $\sum_k [il, k]\,\Delta_{rk}/a$ is written $\{il, r\}$. Δ_{pq}/a is $a^{(pq)}$ (see p. 20).

$[gh, k]$ and $\{gh, k\}$ are called Christoffel's three-index symbols of the first and second kinds respectively. The expression

$$\frac{\partial}{\partial x_i}[gh, k] - \frac{\partial}{\partial x_h}[gi, k] + \sum_p (\{gi, p\}[hk, h] - \{gh, p\}[ik, p])$$

is written $(gkhi)$ and is called sometimes Christoffel's four-index symbol and sometimes a Riemann symbol. In the case of a quadratic form in two variables, a^2K is the only Riemann symbol. These symbols were discovered by Riemann, independently of Christoffel, in his researches on the generalisation of curvature for manifolds of n dimensions.

It is clear from the definition that $[gh, k] = [hg, k]$, and therefore that also $\{gh, k\} = \{hg, k\}$. Also, the three-index symbols of the first kind are linear functions of the first derivatives of the coefficients of F: conversely these first derivatives may be expressed linearly in terms of the symbols. We have in fact

$$\frac{\partial a_{ik}}{\partial x_l} = [il, k] + [kl, i].$$

It is further to be noticed that just as the symbols of the second kind are expressed as linear functions of those of the first kind, so also those of the first kind may be expressed linearly in terms of those of the second kind. The typical equation is

$$[ik, l] = \sum_{\nu} a_{l\nu} \{ik, \nu\}.$$

12. The Riemann symbols are not all linearly independent. We deduce from the definition that there are among them the relations

$$(gkih) = -(gkhi), \quad (kghi) = -(gkhi),$$
$$(ihkg) = (gkhi), \quad (higk) = (gkhi),$$
$$(gkhi) + (ghik) + (gikh) = 0,$$

and it readily follows that there are only $\frac{1}{12}n^2(n^2-1)$ independent Riemann symbols.

13. We now pass on to consider the case of two general quadratic forms F and F'. Symbols derived from the form F' are distinguished by means of accents. If F can be transformed into F' we have the relation $F = F'$. This is equivalent to $\frac{1}{2}n(n+1)$ differential equations of the first order for the x's as functions of the y's, of which a typical one is

$$\sum_{r,s} a_{rs} \frac{\partial x_r}{\partial y_\alpha} \frac{\partial x_s}{\partial y_\beta} = a'_{\alpha\beta} \quad \text{......................(1)}.$$

From these by differentiation are obtained $\frac{1}{2}n^2(n+1)$ equations for second derivatives. Suppose these to be solved for the second derivatives: we get a set of equations of the type

$$\frac{\partial^2 x_r}{\partial y_\alpha \partial y_\beta} + \sum_{i,k} \{ik, r\} \frac{\partial x_i}{\partial y_\alpha} \frac{\partial x_k}{\partial y_\beta} = \sum_{\lambda} \{\alpha\beta, \lambda\}' \frac{\partial x_r}{\partial y_\lambda} \text{...........(2)}.$$

As in the particular case first considered, the number of equations of this type is exactly equal to the number of second derivatives. When, however, the set (1) is differentiated twice we obtain $\frac{1}{4}n^2(n+1)^2$ equations involving third derivatives. There are only $\frac{1}{6}n^2(n+1)(n+2)$ third derivatives of n functions of n variables, and thus by elimination are obtained $\frac{1}{4}n^2(n+1)^2 - \frac{1}{6}n^2(n+1)(n+2) = \frac{1}{12}n^2(n^2-1)$ new equations not involving third derivatives. These may be reduced, by means of the equations (2), to a set involving first derivatives only. Similarly from the equations for fourth derivatives may be deduced a new set of equations in first derivatives only, and so on for all higher derivatives.

We call the set obtained from the equations involving rth derivatives the $(r-1)$th set. The exception to this is the set (1), which is the first set. This is correct because there are no equations of the first order arising from the equations involving second derivatives. The number of equations in the $(r-1)$th set is

$$\tfrac{1}{2}n(r-1)\frac{(n+r-1)!}{(n-2)!\,(r+1)!}.$$

The second set may be got directly from the equations of type (2). If we call the one given $(\alpha\beta)$, then

$$\frac{\partial}{\partial y_\gamma}(\alpha\beta)=0, \quad \frac{\partial}{\partial y_\beta}(\alpha\gamma)=0, \quad \text{and hence } \frac{\partial}{\partial y_\gamma}(\alpha\beta)-\frac{\partial}{\partial y_\beta}(\alpha\gamma)=0.$$

This last equation does not involve third derivatives and may be reduced by means of the equations (2) so as to be of the first order. The totality of new first order equations thus obtained are, after some algebraic modifications, reduced to the form

$$(\alpha\delta\beta\gamma)' = \sum_{g,h,i,k}(gkhi)\frac{\partial x_g}{\partial y_\alpha}\frac{\partial x_h}{\partial y_\beta}\frac{\partial x_i}{\partial y_\gamma}\frac{\partial x_k}{\partial y_\delta} \quad \text{............(3)},$$

where α, β, γ, δ take all values from 1 to n. These constitute the second set. As the number of linearly independent Riemann symbols is $\frac{1}{12}n^2(n^2-1)$, the number of equations in this set is also $\frac{1}{12}n^2(n^2-1)$, as it should be.

A simplification may now be introduced into the calculation of the remaining sets, for it may be seen without much difficulty that the third set may be obtained by differentiating the second set and eliminating the second derivatives involved by means of the equations of type (2), and in general the rth set may be obtained from the $(r-1)$th in exactly the same manner.

14. The quadrilinear form G_4.

We notice that the equations (1) and (3) are similar in form, and just as the former are the conditions for the equivalence of two quadratic differential forms, so the latter may be regarded as the conditions for the equivalence of two differential forms of the fourth order. Let there be four sets of differentials $d^{(1)}y$, $d^{(2)}y$, $d^{(3)}y$, $d^{(4)}y$ of the variables y, and let the corresponding differentials of the variables x be $d^{(1)}x$, $d^{(2)}x$, $d^{(3)}x$, $d^{(4)}x$. Then if

$$G_4 \equiv \sum_{g, k, h, i} (gkhi)\, d^{(1)}x_g d^{(2)}x_k d^{(3)}x_h d^{(4)}x_i,$$

the relations (3) are equivalent to the single equation $G_4' = G_4$.

In this case we are compelled to take four different sets of differentials, for all the equations of (3) could not be obtained from a form in which any pair of the four were made equal to each other. In particular, for example, the form $\Sigma\, (gkhi)\, dx_g dx_k dx_h dx_i$ vanishes identically.

Thus the second set of equations are the conditions for the equivalence of two quadrilinear forms. It will appear that this result is general, in other words that the rth set arises from the equivalence of two $(r+2)$ply linear differential forms.

15. In fact let G_μ be any μ-ply linear differential form, and G_μ' its transformed. If the general coefficient of G_μ is $(i_1 i_2 \dots i_\mu)$ then the equivalence of G_μ and G_μ' leads to a set of relations of type

$$(\alpha_1 \alpha_2 \dots \alpha_\mu)' = \sum_{i_1, \dots, i_\mu} (i_1 i_2 \dots i_\mu) \frac{\partial x_{i_1}}{\partial y_{\alpha_1}} \frac{\partial x_{i_2}}{\partial y_{\alpha_2}} \cdots\cdots \frac{\partial x_{i_\mu}}{\partial y_{\alpha_\mu}} \quad \dots\dots\dots(4)$$

which are of the same form as (3).

Differentiate (4) with respect to y_α and substitute from (2) for the second derivatives of the x's. After a little reduction the equation becomes of exactly the same form as (4) except that μ is changed into $\mu+1$. The new equations obtained may thus be regarded as conditions for the equivalence of two $(\mu+1)$ply linear forms $G_{\mu+1}$ and $G'_{\mu+1}$ The relations which connect the coefficients of $G_{\mu+1}$ with those of G_μ are of type

$$(i_1 i_2 \dots i_\mu i) = \frac{\partial}{\partial x_i} (i_1 i_2 \dots i_\mu) - \sum_\lambda \Big[\{i i_1, \lambda\} (\lambda i_2 \dots i_\mu) + \{i i_2, \lambda\} (i_1 \lambda i_3 \dots i_\mu) + \dots \Big] \dots(5)$$

where there are μ terms in the square bracket on the right, λ replacing each of the letters $i_1, i_2, \dots, i_\mu$ in turn.

A form such as G_4 is called a covariant form of the original form

F, and we now see that from any covariant form of order μ there may be derived another covariant form of order $\mu+1$. Further, the equivalence of the two quadratic forms F and F' leads to the equivalence of a series of covariant forms G_4, G_5, ... derived from F, with the corresponding sequence derived from F'.

16. Sufficiency of the conditions obtained for the equivalence of two forms.

Now suppose that for given initial values of the variables y we can find initial values of the variables x and of the first derivatives $\frac{\partial x}{\partial y}$ which make these two sequences equivalent and also make $F=F'$. Then from the equations (2) we can find uniquely the initial values of the second derivatives, and similarly we can find the values of the third and higher derivatives. There are no contradictions, since in addition to the relations given by the equivalence of the two sequences F, G_4, G_5, ... and F', G_4', G_5', ... there are only enough equations exactly to determine all the higher derivatives. Hence we can find for each of the x's its initial value and the initial values of all its derivatives for a given set of initial values of the y's, and it follows that the differential equations (1) can be formally satisfied, and therefore that the transformation of F into F' is possible. We thus have the important result:

The necessary and sufficient conditions in order that it shall be possible to transform a quadratic form F into another quadratic form F' are that the equations in the variables x, y, $\frac{\partial x}{\partial y}$, derived from the equivalence of the two sequences F, G_4, G_5, ... and F', G_4', G_5', ... shall be algebraically compatible.

We shall now prove this result in another way, and incidentally show how the finite equations of the transformation may be obtained. Let the quantities $\frac{\partial x_i}{\partial y_\alpha}$ be denoted by u_α^i, and consider a linear transformation between two sets of variables X and Y given by the scheme

$$X_i = \sum_{\alpha=1}^{n} u_\alpha^i Y_\alpha. \quad (i=1, \dots, n) \quad \dots\dots\dots\dots\dots\dots(6).$$

Now suppose that different sets of variables $Y^{(1)}$, $Y^{(2)}$, etc. are taken and let the variables X obtained from them by the above transformation be denoted by $X^{(1)}$, $X^{(2)}$, etc. Let G_4 denote

$$\Sigma\,(gkhi)\, X_g^{(1)} X_k^{(2)} X_h^{(3)} X_i^{(4)},$$

then G_4' is the corresponding expression in the variables Y. It is now clear that the equations obtained in the quantities x, y, $\frac{\partial x}{\partial y}$, are precisely the conditions that the sequence F, G_4, $G_5, \ldots$ may transform algebraically into the sequence F', G_4', G_5', ... by means of the scheme (6). These are necessary conditions for the equivalence of the differential forms F and F', but they may not be sufficient, for it may happen that the coefficients u of a possible transformation obtained cannot be regarded as first derivatives of n functions x with respect to n functions y.

For example consider $dx_1^2+dx_2^2=dy_1^2+dy_2^2$. In this case G_4, G_5, etc., and the corresponding forms, vanish identically. Hence the two sequences can be transformed into each other by any linear transformation

$$X_1=u_1 Y_1+u_2 Y_2, \quad X_2=v_1 Y_1+v_2 Y_2$$

which transforms $X_1^2+X_2^2$ into $Y_1^2+Y_2^2$. Such a transformation is given by

$$u_1=\cos\theta, \quad u_2=\sin\theta, \quad v_1=\sin\theta, \quad v_2=-\cos\theta,$$

where θ is any angle whatever. But if the differential forms are to be transformed into each other we must have

$$u_1=\frac{\partial x_1}{\partial y_1}, \quad u_2=\frac{\partial x_1}{\partial y_2}, \quad v_1=\frac{\partial x_2}{\partial y_1}, \quad v_2=\frac{\partial x_2}{\partial y_2}.$$

Hence we must have

$$\frac{\partial u_1}{\partial y_2}=\frac{\partial u_2}{\partial y_1}, \quad \frac{\partial v_1}{\partial y_2}=\frac{\partial v_2}{\partial y_1},$$

and these two conditions are not given by the algebraic equivalence of the two sets of forms. They may easily be seen to impose on θ the additional condition that it must be an absolute constant.

17. Finiteness of the number of conditions for equivalence of two forms.

There are thus certain integrability conditions of the type

$$\frac{\partial u_\alpha}{\partial y_\beta}=\frac{\partial u_\beta}{\partial y_\alpha}$$

to be satisfied by the coefficients u. We shall now prove that *there exists a finite number q such that, if the forms F, $G_4, \ldots, G_q$ are equivalent algebraically to F', $G_4', \ldots, G_q'$, then the integrability conditions can be satisfied and the transformation is possible.*

Assume a linear transformation (6) and let it transform G_μ into G_μ'. We must have

$$(a_1 \ldots a_\mu)' = \sum_{i_1, \ldots, i_\mu} (i_1 \ldots i_\mu)\, u_{\alpha_1}^{i_1} \ldots u_{\alpha_\mu}^{i_\mu}.$$

These may be regarded as algebraic equations for the variables

u, x, n^2+n in number, as functions of the y's. (F may be taken to be G_2 and it may easily be proved that G_3 derived from it is identically zero.) If these equations cannot co-exist, this will appear after a finite number of forms G have been considered. There are two other possibilities:

(1) An order $q-1$ can be determined such that

$$F=F', \quad G_4=G_4', \ldots, G_{q-1}=G'_{q-1}$$

determine the u's and x's as functions of the y's, and these values make $G_q=G_q'$ an identity.

(2) The relations up to order $q-1$ give only p independent equations for the u's and x's, where p is less than n^2+n, and if any values whatever of these variables be taken which satisfy the relations up to order $q-1$, they also satisfy those of order q.

In the first case the relations of order up to $q-1$ give n^2+n independent equations for the u's and x's. Let

$$f_k(u, x, y)=0, \quad (k=1, 2, \ldots, n^2+n),$$

be these relations. Since they determine the u's and x's, their Jacobian with respect to these variables does not vanish.

Write

$$\frac{\partial x_i}{\partial y_a}-u_a^i \equiv \binom{i}{a}, \quad \frac{\partial u_\beta^i}{\partial y_a}+\sum_{r,s}\{rs, i\}\, u_a^r u_\beta^s-\sum_p\{a\beta, p\}' u_p^i \equiv \binom{i}{a\beta} \ldots (7),$$

and suppose that $f_k=0$ is obtained from the relation $G_\lambda=G_\lambda'$. Then after differentiating f_k with respect to y_a, substituting the values of the derivatives from the equations just given, and using a relation from the set given by $G_{\lambda+1}=G'_{\lambda+1}$ we obtain an equation

$$\sum_i \frac{\partial f_k}{\partial x_i}\binom{i}{a}+\sum_{i\beta}\binom{i}{a\beta}\frac{\partial f_k}{\partial u_\beta^i}=0 \quad \ldots\ldots\ldots\ldots\ldots\ldots (8).$$

By giving k all its possible values and keeping a fixed we thus have a set of homogeneous linear equations for the n^2+n quantities $\binom{i}{a}$, $\binom{i}{a\beta}$ obtained by giving i and β all possible values from 1 to n. The determinant of the coefficients of these equations is the Jacobian just mentioned, and therefore it does not vanish. Hence all the quantities $\binom{i}{a}$, $\binom{i}{a\beta}$ vanish and therefore the integrability conditions follow from the algebraic equivalence of the two sequences. In this case, therefore, the transformation of F into F' is possible in only one way, the equations of transformation are obtained from the algebraic equivalence of the two sequences, and the integrability conditions are satisfied of themselves.

In the second case $n^2+n-p=h$, say, of the functions u, x may be chosen arbitrarily, and the others are determinate functions of these. Call the arbitrary ones $Z_1, Z_2, \ldots, Z_h$, and the remainder of the u's and x's $Z_{h+1}, \ldots, Z_{h+p}$, where of course $h+p=n^2+n$. The equations (7) are all of the form

$$\frac{\partial Z_\lambda}{\partial y_\alpha} - \psi_{\lambda\alpha}(Z, y) = (\lambda\alpha) \quad \ldots\ldots\ldots\ldots(9),$$

where λ takes all values from 1 to n^2+n, the ψ's are known functions of their arguments, and $(\lambda\alpha)$ is a quantity $\binom{i}{\alpha}$ or $\binom{i}{\alpha\beta}$.

Let the relations whereby Z_{h+1}, etc. are determined in terms of $Z_1, \ldots, Z_h$ be

$$Z_{h+r} = \phi_r(Z_1, \ldots, Z_h, y) \quad \ldots\ldots\ldots\ldots(10),$$

then the ϕ's are known functions of their arguments, and the equation (8) becomes in this case

$$(h+r, \alpha) = \sum_{t=1}^{h} \frac{\partial \phi_r}{\partial Z_t}(t\alpha) \quad \ldots\ldots\ldots\ldots(11).$$

The equations (7) or (9) may be regarded as differential equations of the first order for the Z's, the quantities $(\lambda\alpha)$ being supposed given. If we write down the conditions of coexistence these are seen to be of the form

$$\frac{\partial(\lambda\alpha)}{\partial y_\beta} - \frac{\partial(\lambda\beta)}{\partial y_\alpha} = \sum_{\lambda\alpha} A_{\lambda\alpha}(\lambda\alpha),$$

where the A's are certain functions of the Z's and y's. By means of (11) these may be turned into equations of the same type, involving only however, the quantities $(\lambda\alpha)$ where λ takes values from 1 to h.

The equations (9) may be reduced to two sets by means of (10); one is a set

$$\frac{\partial Z_\lambda}{\partial y_\alpha} - \chi_{\lambda\alpha}(Z_1, \ldots, Z_h, y) = (\lambda\alpha), \; (\lambda = 1, 2, \ldots, h) \quad \ldots\ldots(12),$$

and the other may be seen to vanish identically in virtue of (10) and (11).

Now the conditions of coexistence of (12) may easily be seen to be the same as those for the set (9), and these are seen to be satisfied if all the quantities $(\lambda\alpha)$ are zero, $(\lambda = 1, \ldots, h)$. It follows from (11) that in this case all the remaining quantities $(\lambda\alpha)$ vanish, and the Z's left arbitrary by the algebraic conditions must be determined to satisfy the equations obtained from (12) by making all the right-hand members zero. These equations possess solutions involving h arbitrary constants, and we see that in this case the transformation of F into F' is possible in ∞^h different ways.

18. Connection of differential with algebraic invariants.

In consequence of the theorems just proved, the problem of the equivalence of two quadratic differential forms is reduced to that of the equivalence of two sets of algebraic forms, where one set is obtained from the other by a linear transformation. The necessary and sufficient condition that it may be possible thus to transform one set of forms into another is that the algebraic invariants of the one set shall be equal to those of the other. It is convenient to extend our definition so as to include relative invariants; a relative invariant is an expression which, under a transformation, repeats itself multiplied by some factor which depends only on the transformation. Let I, I_1, etc. denote a complete system of relative algebraic invariants for the first set, and I', I_1', etc. the corresponding complete system for the second set; we have $I' = kI$, $I_1' = k_1 I_1$, etc. and it is a known theorem that the quantities k are all powers of the determinant of the linear transformation. But this determinant is the Jacobian of the transformation performed on the variables x; it therefore follows that *the invariants $I, I_1, \ldots$ of the algebraic forms $F, G_4, \ldots$ are a complete system of relative differential invariants for the quadratic differential form F, and if under any transformation such an invariant I becomes kI, then k is some power of the Jacobian of the transformation.*

If we take account of differential invariants which involve the magnitudes dx themselves, covariants we may call them, it is clear that they correspond exactly to the covariants of the algebraic forms $F, G_4, \ldots$.

19. In the case where the equations do not determine the transformation of F into F' uniquely, it is easy to see that F must be transformable into itself, for since two different transformations give F' from F, the first of these followed by the inverse of the second gives a transformation of F into itself. Such a case arises, for example, when F is ds^2 for a surface of revolution in space of three dimensions. It is clear that the conditions for this to be possible are expressed by the identical vanishing of certain of the invariants already obtained.

20. Differential parameters.

Now suppose in the general case that we wish to obtain all the invariants when account is taken of systems of functions $f(x_1, \ldots, x_n)$ associated with the quadratic form. We have

$$\frac{\partial f}{\partial y_r} = \sum_i \frac{\partial f}{\partial x_i} \frac{\partial x_i}{\partial y_r};$$

hence any invariant which involves only first derivates of f is taken account of by adding to the set of algebraic forms F, G, the linear form $\sum_i f_i X_i$, where $f_i = \frac{\partial f}{\partial x_i}$.

For second derivatives the case is not so simple. We have, in fact,

$$\frac{\partial^2 f}{\partial y_r \partial y_s} = \sum_{p,q} \frac{\partial^2 f}{\partial x_p \partial x_q} \frac{\partial x_p}{\partial y_r} \frac{\partial x_q}{\partial y_s} + \sum_p \frac{\partial f}{\partial x_p} \frac{\partial^2 x_p}{\partial y_r \partial y_s},$$

and the second term on the right shows that the second derivatives cannot by themselves be taken account of by means of a linear transformation. If the form F is used, however, there are the relations (2) for second derivatives of the x's with respect to the y's, and by means of these equations we have immediately

$$f_{rs}' = \sum_{p,q} f_{pq} \frac{\partial x_p}{\partial y_r} \frac{\partial x_q}{\partial y_s},$$

where
$$f_{pq} = \frac{\partial^2 f}{\partial x_p \partial x_q} - \sum_\lambda \{pq, \lambda\} \frac{\partial f}{\partial x_\lambda}.$$

Now f_{pq} differs from the corresponding second derivative of f by terms involving only first derivatives of f, and it follows that any function, and in particular any invariant, which involves only first and second derivatives of f, may be expressed as a function of the quantities f_p, f_{pq}. But it is clear from the equations of transformation of the quantities f_{pq}, that any invariant involving them may be taken account of by adding the quadratic $\sum_{p,q} f_{pq} X_p X_q'$ to our set of algebraic forms. The extension is immediate, and just as the coefficients of $G_{\mu+1}$ were obtained from those of G_μ (see equation (5)), so may the coefficients f_{pqr} of a cubic form be derived from those of the quadratic; any invariant involving only first, second, and third derivatives of f may be expressed in terms of f_p, f_{pq}, f_{pqr}, and is then seen to be an algebraic invariant of the forms F, G, and the three forms

$$U_1 = \sum_p f_p X_p, \quad U_2 = \sum_{p,q} f_{pq} X_p X_q', \quad U_3 = \sum_{p,q,r} f_{pqr} X_p X_q' X_r''.$$

Generally, any invariant which involves derivatives up to the rth of f is an algebraic invariant of the forms $F, G_4, \ldots, U_1, U_2, \ldots, U_r$, where the coefficients of the successive forms U are calculated in exactly the same way as are the coefficients of the successive forms G.

21. The Absolute Differential Calculus.

These ideas are at the base of the "Absolute Differential Calculus" of Ricci and Levi-Civita. A system of functions $X_{r_1 r_2 \ldots r_m}$ $(r_1, \ldots, r_m = 1, \ldots, n)$ is said to be *covariant* if the transformed system Y is given by

$$Y_{r_1 r_2 \ldots r_m} = \sum_{s_1, s_2, \ldots, s_m} X_{s_1 s_2 \ldots s_m} \frac{\partial x_{s_1}}{\partial y_{r_1}} \frac{\partial x_{s_2}}{\partial y_{r_2}} \cdots \frac{\partial x_{s_m}}{\partial y_{r_m}},$$

and the notation $X_{r_1 \ldots r_m}$ is always used to denote an element of a covariant system.

A *contravariant* system, an element of which is denoted by $X^{(r_1 \ldots r_m)}$, is defined as one which has the transformation scheme

$$Y^{(r_1 \ldots r_m)} = \sum_{s_1, \ldots, s_m} X^{(s_1 \ldots s_m)} \frac{\partial y_{r_1}}{\partial x_{s_1}} \frac{\partial y_{r_2}}{\partial x_{s_2}} \cdots \frac{\partial y_{r_m}}{\partial x_{s_m}}.$$

If X is any function of the variables x, and Y the same function expressed in the variables y, the equation $Y = X$ shows that X may be regarded as belonging either to a covariant or to a contravariant system of order zero.

Since the differentials dy satisfy the equations $dy_r = \sum_s dx_s \frac{\partial y_r}{\partial x_s}$, the differentials dx form a contravariant system of unit order. The coefficients a_{rs} of the fundamental quadratic form F are an example of a covariant system of order two, and if the magnitudes $a^{(pq)}$ are the coefficients of the form reciprocal to F (see p. 10), they are a contravariant system of order two. The laws of composition of systems are very simple:

(1) *Addition.* If $X_{r_1 \ldots r_m}$, $\Xi_{r_1 \ldots r_m}$, are two covariant systems of the same order m, the system $X_{r_1 \ldots r_m} + \Xi_{r_1 \ldots r_m}$ is covariant of the same order. This result holds also for contravariants.

(2) *Multiplication.* If $X_{r_1 \ldots r_m}$, $\Xi_{s_1 \ldots s_p}$, are two covariant systems of orders m and p, the system $X_{r_1 \ldots r_m} \Xi_{s_1 \ldots s_p}$ is covariant of order $m + p$. This theorem also is true of contravariants.

(3) *Composition.* If $X_{r_1 \ldots r_m s_1 \ldots s_p}$ is a covariant system of order $m + p$, and $\Xi^{(s_1 \ldots s_p)}$ is contravariant of order p, the system

$$\sum_{s_1 \ldots, s_p} \Xi^{(s_1 \ldots s_p)} X_{r_1 \ldots r_m s_1 \ldots s_p}$$

is covariant of order m. Similarly the system

$$\sum_{s_1, \dots, s_p} \Xi_{s_1 \dots s_p} X^{(r_1 \dots r_m s_1 \dots s_p)}$$

is contravariant of order m. In particular, if $m = 0$, we have an invariant in either case, so that from two systems of opposite nature and of the same order an invariant is derived by composition.

An example of composition is given by the relation

$$a^{(rs)} = \sum_{p, q} a^{(rp)} a^{(sq)} a_{pq}$$

between the coefficients of the fundamental form F and its reciprocal. Again, by means of the fundamental form, we can derive from any covariant system $X_{r_1 \dots r_m}$ a contravariant system $X^{(r_1 \dots r_m)}$ defined by the equations

$$X^{(r_1 \dots r_m)} = \sum_{s_1, \dots, s_m} a^{(r_1 s_1)} a^{(r_2 s_2)} \dots a^{(r_m s_m)} X_{s_1 \dots s_m},$$

and similarly from a contravariant system $\Xi^{(r_1 \dots r_m)}$ we have

$$\Xi_{r_1 \dots r_m} = \sum_{s_1, \dots, s_m} a_{r_1 s_1} a_{r_2 s_2} \dots a_{r_m s_m} \Xi^{(s_1 \dots s_m)}.$$

It is easily seen that if $X^{(r_1 \dots r_m)}$ is the system derived from $X_{r_1 \dots r_m}$, then $X_{r_1 \dots r_m}$ is the system derived from $X^{(r_1 \dots r_m)}$. For this reason $X_{r_1 \dots r_m}$ and $X^{(r_1 \dots r_m)}$ are said to be reciprocal with respect to the fundamental form F.

From the two equations just given it readily follows that

$$\sum_{r_1, \dots, r_m} X_{r_1 \dots r_m} \Xi^{(r_1 \dots r_m)} = \sum_{r_1, \dots, r_m} X^{(r_1 \dots r_m)} \Xi_{r_1 \dots r_m},$$

or *every invariant composed of a covariant system and a contravariant system of the same order is the same as the invariant composed of their reciprocals.*

22. The system ϵ.

There is a certain system of order n of much importance in the theory. Let a denote the discriminant of the form F, and give $\sqrt{a}$ a definite sign for some given set of variables x; make the convention that this sign does not change when the transformation is made to a new set of variables y, provided the Jacobian of the x's with respect to the y's is positive. If however the Jacobian is negative let the sign of this square root be changed. The system $\epsilon_{r_1 \dots r_n}$ is now defined to be zero unless all the r's are different, and, the r's being all different, is equal to $+\sqrt{a}$ or to $-\sqrt{a}$ according as the class of the permutation $(r_1 r_2 \dots r_m)$ is even or odd with respect to $(1, 2, \dots, n)$. This system is covariant, and the elements of the reciprocal system are equal to zero or to $\pm \frac{1}{\sqrt{a}}$.

If $\Delta (z_1 \ldots z_n)$ is the Jacobian divided by $\sqrt{a}$ of n functions $z_1, \ldots, z_n$ with respect to the x's we have the identity

$$\Delta (z_1 \ldots z_n) = \sum_{r_1, \ldots, r_n} \epsilon^{(r_1 \ldots r_n)} \frac{\partial z_1}{\partial x_{r_1}} \frac{\partial z_2}{\partial x_{r_2}} \cdots \frac{\partial z_n}{\partial x_{r_n}},$$

which shows that Δ is invariant, and at the same time renders applicable to it the processes of the Absolute Differential Calculus.

23. Covariant and contravariant differentiation.

From a covariant system of order m may be derived by the method of Christoffel a system of order $m+1$. With the present notation equation (5) becomes

$$X_{r_1 \ldots r_m r_{m+1}} = \frac{\partial}{\partial x_{m+1}} (X_{r_1 \ldots r_m}) - \sum_{l=1}^{m} \sum_{q} \{r_l r_{m+1}, q\} X_{r_1 \ldots r_{l-1} q r_{l+1} \ldots r_m} \ldots (5'),$$

and this gives the elements of the derived system. The elements of the derived system are called the *covariant derivatives* of the original system.

If we substitute in (5′) for the X's their reciprocals, we have a set of equations connecting the elements of a contravariant system of order $m+1$ with those of a similar system of order m. If these equations are solved for the elements of order $m+1$ they give

$$X^{(r_1 \ldots r_m r_{m+1})} = \sum_{t} a^{(t r_{m+1})} \left\{ \frac{\partial}{\partial x_t} (X^{(r_1 \ldots r_m)}) + \sum_{l=1}^{m} \sum_{q} \{t_q, r_l\} X^{(r_1 \ldots r_{l-1} q r_{l+1} \ldots r_m)} \right\} \ldots\ldots (5''),$$

and we have the equations of what may be called the *contravariant derivatives* of a given system.

It is to be noticed that the derivative of a particular element is indicated by writing an additional suffix or index at the right end of those denoting the original element; for example the covariant derivative of X_{pq} with respect to x_r is X_{pqr}, and this is in general not the same as X_{prq}.

The laws for differentiation of sums and products of systems of the same kind are exactly the same as those for ordinary differentiation. For example, from the equations

$$Z_{pq} = X_{pq} + Y_{pq}, \quad Z_{pqrst} = X_{pq} Y_{rst},$$

it is easy to deduce the relations

$$Z_{pqk} = X_{pqk} + Y_{pqk}, \quad Z_{pqrstk} = X_{pqk} Y_{rst} + X_{pq} Y_{rstk},$$

and similarly for any covariant or contravariant systems.

If $$Z_{r_1 \ldots r_m} = \sum_{s_1, \ldots, s_p} Y^{(s_1 \ldots s_p)} X_{r_1 \ldots r_m s_1 \ldots s_p},$$

the elements of the first derived system $Z_{r_1 \ldots r_m k}$ are readily seen to be given by the equation

$$Z_{r_1 \ldots r_m k} = \sum_{s_1, \ldots, s_p} Y^{(s_1 \ldots s_p)} X_{r_1 \ldots r_m s_1 \ldots s_p k} + \sum_{s_1, \ldots, s_p, t} Y^{(s_1 \ldots s_p t)} a_{tk} X_{r_1 \ldots r_m s_1 \ldots s_p},$$

and there is a corresponding formula for a composite contravariant system.

24. Now let $X_{r_1 \ldots r_m}$ be any covariant system whatever, and form its second derived system. We have the identity

$$X_{r_1 \ldots r_m hk} - X_{r_1 \ldots r_m kh} = \sum_{l=1}^{m} \sum_{p, q} a^{(pq)} a_{hkr_l p} X_{r_1 \ldots r_{l-1} q r_{l+1} \ldots r_m}, \quad \ldots (13),$$

where $a_{hkr_l p}$ is written for the Riemann symbol $(hkr_l p)$. (This notation is justified since the Riemann symbols have been shown to be elements of a covariant system of the fourth order.) It thus appears that the element $X_{r_1 \ldots r_m hk}$ is not in general equal to the element $X_{r_1 \ldots r_m kh}$. In fact, if all covariant differential operators are interchangeable, all the Riemann symbols must vanish identically. If the fundamental form is $\sum_i dx_i^2$, all the Riemann symbols do vanish, and it readily appears from the result of Christoffel already given that if these symbols vanish for a quadratic form, that form must be reducible to the sum of the squares of n perfect differentials. (There are some very particular cases of exception to this, of no importance in our theory.) In this case covariant differentiation reduces to ordinary differentiation.

The number of linearly independent Riemann symbols has been proved to be $\frac{1}{12} n^2 (n^2 - 1)$. In particular for $n = 2$ there is only one such symbol, G say, where $G = (1212)$, and for $n = 3$ there are six symbols.

The six equations obtained by Lamé (see his *Leçons sur les coordonnées curvilignes*) in connection with triply orthogonal systems of surfaces, are got by equating to zero the particular values the Riemann symbols take for the form $Pdx^2 + Qdy^2 + Rdz^2$. These six equations for a general quadratic form in three variables are given by Cayley (*Coll. Math. Papers*, vol. XII. p. 13).

A certain amount of symmetry may be introduced into the case of $n=3$ if it is postulated that we may replace one index by another when their difference is 3. The linearly independent symbols may then all be expressed in the form $a_{r+1,\ r+2,\ s+1,\ s+2}$, and if $a^{(rs)}$ be written for this symbol, the system $a^{(rs)}$ may be shown to be contravariant of the second order. In these two cases the relation between second covariant derivatives may be expressed in the forms:

$$\text{for } n=2, \qquad \sum_{r,s} \epsilon^{(rs)} X_{r_1 \dots r_m rs} = G \sum_{l=1}^{m} \sum_{r,s} a^{(rs)} \epsilon_{rr_l} X_{r_1 \dots r_{l-1} s r_{l+1} \dots r_m},$$

$$\text{for } n=3, \qquad \sum_{s,t} \epsilon^{(rst)} X_{r_1 \dots r_m st} = \sum_{q,s,t} a^{(qs)} a^{(rt)} \sum_{l=1}^{m} \epsilon_{r_l st} X_{r_1 \dots r_{l-1} q r_{l+1} \dots r_m}.$$

25. In the general case we have a quadratic differential form F. Its coefficients form a covariant system of the second order. The system derived from this is identically zero. We have, however, the quadrilinear form G_4, and its coefficients are a covariant system of the fourth order. The covariant derivatives of this system are the coefficients of G_5, and similarly each form G_μ has coefficients which are the covariant derivatives of the coefficients of $G_{\mu-1}$. Again suppose we have any number of functions U, V, etc. associated with F, and suppose that the transformation that changes F into F' changes U into U'. For the equivalence of U with U' it is clear from what has been said, that the algebraic forms U_1, U_2, etc., where $U_{\mu+1}$ is the form obtained by covariant differentiation from U_μ, must be linearly transformed into U_1', U_2', etc. by the transformation which changes the sequence F, G into the sequence F', G'. It may easily be shown that all invariant relations arising from the differential form F and the functions U, V, etc. must be invariant relations under a linear transformation for the system of algebraic forms

$$F, G_4, G_5, \dots, U_1, U_2, \dots, V_1, V_2, \dots, \text{etc.},$$

which we call the set S. In particular, all differential invariants of F, U, V, etc. must be algebraic invariants of the set S. We notice that the coefficients of any form in S are a covariant system, and that all the forms except G_4 are obtained by covariant differentiation. Further, if C be any algebraic covariant of S, it is clear that the coefficients of C are a covariant system.

From a covariant system of order r and a contravariant system of order s there can be constructed by composition a covariant system of order $r-s$ or a contravariant system of order $s-r$ according as r is greater than or less than s. If in particular $r = s$ we have an invariant.

Initially we have the covariant systems given by the coefficients of the set S, and the contravariant system ϵ. From these we obtain new covariant and contravariant systems by composition, and, in particular, invariants are thus obtained. By repeated application of the principle of composition new systems may be obtained, and it may be proved that all covariant and contravariant systems, and all invariants, may be thus obtained. The theorem is one on invariants and covariants of algebraic forms and is not difficult to prove.

26. Application to the theory of invariants of binary forms.

The method of proof will be illustrated for the case of a single binary form, though the process is perfectly general. We recall the ordinary symbolic notation for algebraic invariants and covariants. Let $a_0 x^n + n a_1 x^{n-1} y + \ldots + a_n y^n$ be written symbolically

$$(\alpha_1 x + \alpha_2 y)^n,$$

where the symbols $\alpha_1 \alpha_2$ have no meaning by themselves, but $\alpha_1{}^n = a_0$, $\alpha_1{}^{n-1} \alpha_2 = a_1$, etc. Then any expression involving the a's may be expressed in terms of the α's. If, however, we have a term, $a_0 a_2$ for example, it becomes in terms of the α's $\alpha_1{}^{2n-2} \alpha_2{}^2$, and this might equally be the term $a_1{}^2$. To avoid ambiguities of this kind we must arrange to have no expression of other than the nth degree in α_1 and α_2 combined, and we must therefore introduce equivalent symbols. Let there be introduced equivalent sets of symbols $\alpha_1 \alpha_2$, $\beta_1 \beta_2$, etc. such that the form is symbolically

$$(\alpha_1 x + \alpha_2 y)^n \equiv (\beta_1 x + \beta_2 y)^n \equiv \text{etc.}$$

Then $a_0 a_2$ is $\alpha_1{}^n \beta_1{}^{n-2} \beta_2{}^2$, and similarly, if enough equivalent symbols are used, all products of the coefficients of the form may be expressed without ambiguity. If the determinant $\alpha_1 \beta_2 - \alpha_2 \beta_1$ is denoted by $(\alpha\beta)$ it is easily seen that $(\alpha\beta)$, though without actual meaning by itself, satisfies the conditions for an algebraic invariant of the binary form. Hence products of such determinants satisfy these conditions, and therefore, if the product has an actual interpretation in terms of the a's, it is an invariant. For example, $(\alpha\beta)^n$ is such an invariant, and if $n = 4$, $(\beta\gamma)^2 (\gamma\alpha)^2 (\alpha\beta)^2$ is an invariant. It is also obvious that $\alpha_1 x + \alpha_2 y$ is a symbolic covariant, and hence if this expression be written α_x, any product of factors such as $(\alpha\beta)$, α_x is a covariant, symbolic usually, but with an actual interpretation if it is of the nth degree in each of the sets of symbols α, β, γ, etc. Further, it is capable of proof that any invariant or covariant whatever of the binary form may be expressed as a sum of products of factors of the types $(\alpha\beta)$, α_x.

By a slight extension of the symbolic notation any form such as

$$a_0 x_1 x_2 + a_1 x_1 y_2 + a_2 x_2 y_1 + a_3 y_1 y_2,$$

linear in sets of cogredient variables $x_1 y_1$, $x_2 y_2$, etc. may be symbolically expressed as $(\alpha_1 x_1 + \alpha_2 y_1)(\alpha_1' x_2 + \alpha_2' y_2)$, etc., where α, α', are symbols to be interpreted as before. (In our particular case, for example, $\alpha_1 \alpha_1' = a_0$, $\alpha_1 \alpha_2' = a_1$, etc.) It may be proved that any invariant is a sum of products of factors such as $(\alpha\alpha')$, $(\alpha\beta)$, α_x, etc.

Now let a base system of variables ξ_t, η_t be introduced, and let J_x denote the determinant of the coefficients whereby the variables x_t, y_t are derived from this set. Then if x_t, y_t are transformed into x_t', y_t' by any transformation,

$$J_{x'} = J_x J \begin{pmatrix} x_t' \, y_t' \\ x_t \, y_t \end{pmatrix}.$$

Next introduce a system ϵ_{rs} such that $\epsilon_{rs} = 0$ if $r = s$, and is equal to $\pm 1/J_x$ if $r \neq s$, the sign being determined as before. Similarly introduce a system $\epsilon^{(rs)} = 0$ or $\pm J_x$. Then it is easily seen that the former system is covariant, the latter contravariant.

Now $$(\alpha\beta) \equiv (\alpha_1 \beta_2 - \alpha_2 \beta_1) \equiv \frac{1}{J_x} \sum_{rs} \epsilon^{(rs)} \alpha_r \beta_s,$$

and $$\alpha_x = \alpha_1 x + \alpha_2 y.$$

The system x, y satisfies the definition of a contravariant system of the first order. (The terms covariant, contravariant, applied to systems in the absolute calculus, are retained for the algebraic theory, though no special meaning is now attached to them.) Hence, if we change our notation, we may call them $x^{(1)}$, $x^{(2)}$, and

$$\alpha_x = \sum_r \alpha_r x^{(r)}.$$

Thus any invariant or covariant consisting of a product of factors of the types $(\alpha\beta)$, α_x, is clearly obtained by composition from the systems $\epsilon^{(rs)}$, α_r, $x^{(r)}$, and when the products of α's are replaced by their actual values in terms of the a's the theorem is evident.

For example, consider the invariant of a single quadratic

$$\sum_{rs} a_{rs} x^{(r)} x^{(s)} = (\sum_r \alpha_r x^{(r)})^2.$$

The invariant in symbolic notation is

$$(\alpha\beta)^2 \equiv \frac{1}{J_x^2} \sum_{rs} \epsilon^{(rs)} \alpha_r \beta_s \times \sum_{pq} \epsilon^{(pq)} \alpha_p \beta_q$$

$$\equiv \frac{1}{J_x^2} \sum_{rspq} \epsilon^{(rs)} \epsilon^{(pq)} \alpha_r \alpha_p \beta_s \beta_q$$

$$\equiv \frac{1}{J_x^2} \sum_{rspq} \epsilon^{(rs)} \epsilon^{(pq)} a_{rp} a_{sq}.$$

The factor J_x^2 and the base system of variables have been introduced because symbolic algebraic invariants and covariants are merely relative, whilst those of the absolute calculus are absolute.

It is now clear, generally, that all the covariant and contravariant systems arising from given systems may be obtained by composition, and further, we have obtained a useful result in connection with algebraic invariants. This result is that *if the notation for the coefficients of given algebraic forms is properly chosen, any invariant or covariant may be expressed non-symbolically in such a form that its invariance is at once in evidence.*

27. To return to the general theory, the result obtained is that all the differential invariants of order $\leqslant \mu$ may be expressed in the form of invariants of the absolute differential calculus, by means of the coefficients of the following forms :

(1) the fundamental quadratic form ;

(2) the associated functions f and their covariant derivatives of orders up to μ ;

(3) a quadrilinear form G_4, and its covariant derivatives of orders up to $\mu - 2$.

28. It is now clear that if any problem is given in which a quadratic differential form is fundamental, it suffices to replace ordinary differentiation by covariant differentiation with respect to the quadratic form to obtain the equations of the problem in an invariant shape. It may happen that we have the equations expressed in a simple form due to the choice of a particular set of variables, and by means of this theory we may express them in terms of a general set of variables without going through the process of calculation.

As an example consider potential functions in space of three dimensions. They satisfy the equation

$$\frac{\partial^2 V}{\partial x^2} + \frac{\partial^2 V}{\partial y^2} + \frac{\partial^2 V}{\partial z^2} = 0,$$

when the coordinates are rectangular cartesian. The quadratic form is now $dx^2+dy^2+dz^2$, and when the equation is written $\Sigma a^{(rs)} V_{rs}=0$, its invariance becomes intuitive and we have its expression in a general set of variables x_1, x_2, x_3. For other examples see a paper by the author, *Bull. Amer. Math. Soc.* (1906), p. 379.

One of the most important properties of a covariant or contravariant system is that if all its elements vanish they do so independently of the particular variables chosen. In other words, the system of equations, *e.g.* $X_{rst}=0$, is an invariant system.

CHAPTER III

THE METHOD OF LIE

29. We now consider the second of the methods used to determine differential invariants. This method depends on the use of infinitesimal transformations, and is originally due to Lie. It will be of use to recall the main points of Lie's theory. This we shall do for the case in which there are only two independent variables, though the ideas are perfectly general.

Suppose that there are two sets of variables x, y, and x', y', connected by relations

$$x' = f(x, y), \quad y' = \phi(x, y).$$

These relations will define a transformation scheme, provided the variables x, y can be determined from them in terms x', y', *i.e.* provided the Jacobian of the functions f, ϕ, does not vanish. The operation of replacing x, y by x', y', in any function V of x, y, may be denoted by SV, that is to say, $SV(x, y) = V(x', y')$. Now let T denote a similar transformation scheme, then by applying first T and then S we obtain a scheme represented by ST.

For example, let $x' = ax + b$, $y' = ay + b$, be the scheme S, and $x' = x^2$, $y' = y^2$ be the scheme T. Also let V be $x + y$. Then

$$x^2 + y^2 = T(x + y), \quad \text{and} \quad (ax + b)^2 + (ay + b)^2 = S(x^2 + y^2) = ST(x + y).$$

Also $$ax + by + 2b = S(x + y),$$

and therefore $$ax^2 + ay^2 + 2b = TS(x + y).$$

It is easily seen from this example that TS and ST are not in general the same, so that it is necessary to take account of the order in which the operations are performed.

Starting therefore with the two schemes S and T, we obtain two more schemes given by ST and TS. We have also the scheme SS, which may be written shortly S^2, and in general we have schemes given by the operations $S^\lambda T^\mu S^\nu \ldots$.

In exactly the same way, if any number of transformations are given, we can obtain others by repetition of these transformations. If it happens that all the transformations that can be obtained in this way are included in the original set, this original set is said to form a Group of transformations. For example, the set of six transformations $x' = x$, $x' = 1 - x$, $x' = 1/x$, $x' = (x-1)/x$, $x' = x/(x-1)$, $x' = 1/(1-x)$, is a group.

Another group is given by all the transformations of the form $x' = ax + by$, $y' = cx + dy$, where a, b, c, d, are arbitrary.

A third example is the set of transformations $x' = f(x)$, $y' = y\phi(x)$ where f and ϕ are arbitrary functions of x.

These three examples illustrate different types of group. The first contains only six operations, and in fact these may be obtained by repetition of the two $x' = 1 - x$, $x' = 1/x$.

The second contains a quadruple infinity of transformations, since it depends on four arbitrary constants. The third depends on two arbitrary functions, and in comparison with the second may be said to depend on an infinite number of arbitrary constants.

A transformation of a group is said to be infinitesimal when the variables x', y' obtained from it differ infinitesimally from the variables x, y. The first of the three examples contains no infinitesimal transformations. The second contains, for example, the one given by $x' = x + \epsilon y$, $y' = y$, where ϵ is infinitesimal, and similarly the third contains infinitesimal transformations.

30. Continuous groups.

The group is said to be continuous if all its transformations may be built up from infinitesimal transformations. The first of the examples given is not continuous, the second and third are continuous groups. We shall be concerned only with continuous groups in future work. (It is to be noticed that a group may be discontinuous and yet contain an infinite number of transformations. For example, the modular group $x' = ax + by$, $y' = cx + dy$, where a, b, c, d are any integers such that $ad - bc = 1$, is discontinuous.)

A continuous group is said to be finite if all its transformations are determined by a finite number of parameters, whereas, if it involves arbitrary functions, it is said to be infinite. The second example given is a finite continuous four parameter group; the third is an infinite continuous group. It may be proved that a finite continuous group which depends on r parameters may be generated from r infinitesimal transformations. The second of our three examples may be generated from the four infinitesimal transformations

$$1\ \begin{cases} x' = x + \alpha y \\ y' = y \end{cases} \quad 2\ \begin{cases} x' = x \\ y' = y + \beta x \end{cases} \quad 3\ \begin{cases} x' = x + \gamma x \\ y' = y \end{cases} \quad 4\ \begin{cases} x' = x \\ y' = y + \delta y \end{cases}$$

where α, β, γ, δ are infinitesimal. The third may be generated from infinitesimal transformations of the type

$$x' = x + \alpha\xi(x), \quad y' = y + \beta y\eta(x),$$

where ξ and η are arbitrary functions of x, and α, β, are infinitesimal.

31. Invariant of a group.

Now suppose that $V(x, y)$ is a function of the variables x, y which is unchanged when any transformation of a certain continuous group is performed upon it. V is said to be an invariant of the group. Since the transformations of the group can all be built up from infinitesimal transformations, it follows that the necessary and sufficient conditions for invariance are that V shall be invariant under all the infinitesimal transformations of the group.

Let $x' = x + \xi dt$, $y' = y + \eta dt$ be any infinitesimal transformation of the group. If V is an invariant we must have

$$V(x + \xi dt, y + \eta dt) = V(x, y).$$

Hence

$$\Omega V \equiv \xi \frac{\partial V}{\partial x} + \eta \frac{\partial V}{\partial y} = 0.$$

By making use of all the infinitesimal transformations of the group, we thus obtain a set of linear partial differential equations which all invariants must satisfy, and it may be proved that this set is complete and that all its solutions are invariants.

32. Extension of a group.

Now suppose that we have a group of transformations on the variables x, y, u, v, and suppose further that x, y, are independent variables, and that u, v, are functions of these. A transformation of the group determines x', y', u', v' as functions of x, y, u, v. By differentiating the equations of transformation, we may obtain equations whereby the derivatives of the transformed variables are determined from the original variables and their derivatives. The totality of all the equations of transformation now obtained may be shown to form a group which is said to be *extended* from the original group.

Suppose further that there are other variables, a, b, c for example, dependent on x, y, u, v, which are such that there are relations of the type

$$f(a, b, c, x, y, u, v) = f(a', b', c', x', y', u', v')$$

connecting their original and their transformed values. Then if these can be solved for a', b', c' we may add their solved expressions to the equations of the group, and thus obtain a further extended group. Consider for example a transformation on two variables x, y, of the type

$$x' = x'(x, y), \quad y' = y'(x, y).$$

We may regard x and y as functions of a variable α which is invariant under our group. Then $\alpha' = \alpha$ and we have

$$\frac{dx'}{d\alpha} = \frac{\partial x'}{\partial x}\frac{dx}{d\alpha} + \frac{\partial x'}{\partial y}\frac{dy}{d\alpha}$$

with a similar expression for $\dfrac{dy'}{d\alpha}$.

We thus have the transformation equations for the derivatives $\dfrac{dx}{d\alpha}$, $\dfrac{dy}{d\alpha}$, and similarly those for higher derivatives may be determined.

Again let a, b, c be three variables, functions of x, y such that

$$a dx^2 + 2b dx dy + c dy^2 = a' dx'^2 + 2b' dx' dy' + c' dy'^2.$$

This relation leads to three equations for a', b', c' in terms of a, b, c, x, y, and we thus have a group in these five variables which may be readily extended so as to give the transformation equations for the derivatives of the variables a, b, c with respect to x and y.

33. Application to determination of invariants of a quadratic form.

The problem of the determination of differential invariants of a quadratic differential form is now seen to be that of the determination of all invariants of a certain extended group.

We start with a general point transformation on the n variables $x_1, \ldots, x_n$ given by the equations

$$x_i' = x_i'(x_1, \ldots, x_n), \qquad (i = 1, 2, \ldots, n).$$

Certain dependent variables f, functions of the x's, are introduced, which satisfy the relations $f' = f$. In addition there are introduced other dependent variables a_{rs}, functions of the x's, of which the transformed expressions are determined from the identity

$$\Sigma a_{rs}' dx_r' dx_s' = \Sigma a_{rs} dx_r dx_s.$$

It is convenient also to introduce as variables the total differentials dx, d^2x, etc. (These last might be introduced by adding an equation $\alpha' = \alpha$ to the equations of transformation, and regarding the x's as functions of α. The total differentials would then be practically the same as the derivatives of the x's with respect to α.) The group of all such point transformations is then extended so as to include the transformation equations for the new variables introduced and their derivatives, and our problem is to determine all the invariants of this extended group. To do this it is necessary to obtain the infinitesimal transformations of the group, and from these we obtain a complete system of linear differential equations the solutions of which are the invariants.

34. The case of two independent variables.

As a first case we consider a quadratic form in two variables x, y. Let

$$x' = x + \xi(x, y)\,\delta t, \quad y' = y + \eta(x, y)\,\delta t$$

be an infinitesimal transformation of the original group, then ξ, η are arbitrary functions of x, y. Let the quadratic form be

$$Edx^2 + 2Fdxdy + Gdy^2,$$

then $$E'dx'^2 + 2F'dx'dy' + G'dy'^2 = Edx^2 + 2Fdxdy + Gdy^2.$$

Hence if $$E' = E + \delta E, \text{ etc.},$$

$$\delta E dx^2 + 2\delta F dxdy + \delta G dy^2 + [2Edxd\xi + 2F(dxd\eta + dyd\xi) + 2Gdyd\eta]\,\delta t = 0,$$

and therefore

$$-\delta E = (2E\xi_x + 2F\eta_x)\,\delta t,$$
$$-\delta F = (E\xi_y + F\xi_x + F\eta_y + G\eta_x)\,\delta t,$$
$$-\delta G = (2F\xi_y + 2G\eta_y)\,\delta t.$$

Also $$\delta dx = (\xi_x dx + \xi_y dy)\,\delta t, \quad \delta dy = (\eta_x dx + \eta_y dy)\,\delta t,$$

and if f is an associated function, $\delta f = 0$.

Now if z be any one of the dependent variables we must have

$$dz - z_x dx - z_y dy = 0,$$

and therefore

$$d\delta z - z_x d\delta x - z_y d\delta y - \delta z_x dx - \delta z_y dy = 0,$$

from which the increments δz_x, δz_y, may be obtained by equating to zero the coefficients of dx and dy, and the process may be repeated for higher derivatives. For instance, the increments of the derivatives of f are given by the equations

$$-\delta f_x = f_x\xi_x + f_y\eta_x, \quad -\delta f_y = f_x\xi_y + f_y\eta_y.$$

The infinitesimal operator of the group in the variables x, y, E, F, G, f, f_x, f_y, is thus

$$\Omega \equiv \xi \frac{\partial}{\partial x} + \eta \frac{\partial}{\partial y} - (2E\xi_x + 2F\eta_x)\frac{\partial}{\partial E} - (E\xi_y + F\xi_x + F\eta_y + G\eta_x)\frac{\partial}{\partial F}$$
$$- (2F\xi_y + 2G\eta_y)\frac{\partial}{\partial G} - (f_x\xi_x + f_y\eta_x)\frac{\partial}{\partial f_x} - (f_x\xi_y + f_y\eta_y)\frac{\partial}{\partial f_y}.$$

Since ξ and η are arbitrary, it is clear that any invariant must satisfy the set of differential equations obtained from the coefficients of ξ, η and their derivatives, in the equation $\Omega I = 0$, where I is an invariant, and hence we have

$$\frac{\partial I}{\partial x} = 0, \quad \frac{\partial I}{\partial y} = 0, \quad 2E\frac{\partial I}{\partial E} + F\frac{\partial I}{\partial F} + f_x\frac{\partial I}{\partial f_x} = 0,$$
$$2G\frac{\partial I}{\partial G} + F\frac{\partial I}{\partial F} + f_y\frac{\partial I}{\partial f_y} = 0, \quad 2F\frac{\partial I}{\partial E} + G\frac{\partial I}{\partial F} + f_y\frac{\partial I}{\partial f_x} = 0,$$
$$2F\frac{\partial I}{\partial G} + E\frac{\partial I}{\partial F} + f_x\frac{\partial I}{\partial f_y} = 0.$$

This is a complete system of six independent linear equations in eight variables, and therefore it possesses two functionally independent solutions. There are thus two invariants in these eight variables ; one is f, and the other is readily found to be Beltrami's first differential parameter

$$\frac{Ef_y^2 - 2Ff_yf_x + Gf_x^2}{EG - F^2}.$$

There are several points to be noted in connection with this example. In the first place, the finite equations of the group are not wanted, and therefore it is only necessary to calculate the increments of the various variables for an infinitesimal transformation. Secondly, the method given for finding the increments of various derivatives would be very long in any general case, and some more convenient method must be used. It is to be noticed also that if the group be only so far extended that no variables have increments containing derivatives higher than the pth of ξ and η, the variables occurring will be

(1) $x, y, dx, dy, \ldots, d^px, d^py$,

(2) the functions f and their derivatives of orders up to p,

(3) the functions E, F, G and their derivatives of orders up to $p-1$.

In this case the number of linear equations is equal to the number of derivatives of ξ and η involved, that is

$$2(1 + 2 + 3 + \ldots + \overline{p+1}) \equiv (p+1)(p+2).$$

The number of variables in the linear system is

$$2(p+1)+\tfrac{1}{2}k(p+1)(p+2)+\tfrac{3}{2}p(p+1),$$

where k is the number of different functions f. Hence if the equations are linearly independent the number of invariants is

$$2(p+1)+\tfrac{1}{2}k(p+1)(p+2)+\tfrac{3}{2}p(p+1)-(p+1)(p+2).$$

Again, there is no need to take account of more than two functions f, for any other function may be expressed in terms of these two, say $f^{(1)}$, $f^{(2)}$. All the derivatives of another function f may therefore be expressed as functions of $f^{(1)}$, $f^{(2)}$ and their derivatives. It follows that any invariant involving derivatives of f may be expressed as an invariant that involves only $f^{(1)}$, $f^{(2)}$ and their derivatives. If, then, $k=2$ and the linear equations are independent, the number of invariants is $\tfrac{1}{2}(3p+4)(p+1)$. If the total differentials (1) do not occur, and if $k=0$, the number is apparently $\tfrac{1}{2}p(p-3)$. In this particular case the equations are, however, not independent. They are connected by one linear relation, and the number of invariants is $\tfrac{1}{2}(p-1)(p-2)$. These invariants involve only E, F, G and their derivatives. They are known as Gaussian invariants, and the first of them is the well-known quantity K.

35. The general case of n independent variables.

We shall now consider the general case of a quadratic form in n variables. Let the variables be $x_1, \ldots, x_n$ and let the infinitesimal transformation be given by

$$\delta x_i = \xi_i(x_1, \ldots, x_n)\,\delta t, \qquad (i=1, \ldots, n)$$

$\xi_i(x_1, \ldots, x_n)$ will be denoted by $\xi_i(x)$. Also let f denote any function of the x's; the increments of the various derivatives of f are needed and to determine them we use the method of Forsyth.

36. Determination of the increments.

Let x_i denote the original, x_i' the transformed variables, and let x_i+k_i become $x_i'+k_i'$ $(i=1, \ldots, n)$. Then

$$k_i' = x_i'+k_i'-x_i' = x_i+k_i+\xi_i(x+k)\,\delta t-[x_i+\xi_i(x)\,\delta t].$$

Hence $k_i' = k_i+[\xi_i(x+k)-\xi_i(x)]\,\delta t$, and therefore if f denote any function of the variables x, and f' its transformed,

$$\begin{aligned} f(x+k) &= f'(x'+k') = f'(x'+k+\overline{\xi(x+k)-\xi(x)}\,\delta t) \\ &= f'(x'+k)+\sum_{i=1}^{n}(\xi_i(x+k)-\xi_i(x))\frac{\partial f'(x'+k)}{\partial(x_i'+k_i)}\,\delta t+\ldots, \end{aligned}$$

or, if small quantities of order higher than the first are neglected,

$$\frac{df(x+k)}{dt} + \sum_{i=1}^{n} (\xi_i(x+k) - \xi_i(x)) \frac{\partial f(x+k)}{\partial (x_i + k_i)} = 0,$$

where d/dt operates only on x and not on k. By equating to zero the coefficients of the various powers of the k's, we have all the increments of the derivatives of f. Exactly as in the particular case of two independent variables, the increments of the quantities a_{rs} may be obtained by comparison of the quadratic form and its transformed, and then Forsyth's method will give the increments of the derivatives of the a's.

The increments of the a's are given by

$$-\frac{da_{rs}}{dt} = \sum_h a_{rh} \frac{\partial \xi_h}{\partial x_s} + \sum_h a_{hs} \frac{\partial \xi_h}{\partial x_r}.$$

Also if small quantities of order higher than the first are neglected,

$$a_{rs}'(x' + k') = a_{rs}'(x' + k + \overline{\xi(x+k) - \xi(x)}\,\delta t)$$

$$= a_{rs}'(x' + k) + \sum_i [\xi_i(x+k) - \xi_i(x)] \frac{\partial a_{rs}(x+k)}{\partial (x_i + k_i)} \delta t,$$

and therefore

$$\frac{d}{dt} a_{rs}(x+k) + \left\{ \sum_h a_{rh}(x+k) \frac{\partial \xi_h(x+k)}{\partial (x_s + k_s)} + \sum_h a_{hs}(x+k) \frac{\partial \xi_h(x+k)}{\partial (x_r + k_r)} \right.$$

$$\left. + \sum_i [\xi_i(x+k) - \xi_i(x)] \frac{\partial a_{rs}(x+k)}{\partial (x_i + k_i)} \right\} = 0,$$

where d/dt does not operate on k, and, as before, the increments of the derivatives of a_{rs} are given by equating to zero the coefficients of the various powers of the k's in this expression.

37. Definition of Rank.

Now suppose that the quadratic form $\Sigma\, a_{rs} dx_r dx_s$ is of such a type that it may be written $\sum_{\lambda=1}^{m} du^2$, where the u's are functions of the x's. It is clear that, unless the discriminant of the form vanishes, a case that we definitely exclude, m must be equal to or greater than n; if m is taken to be $\frac{1}{2}n(n+1)$, the equivalence of the two expressions for the form leads to $\frac{1}{2}n(n+1)$ differential equations of the first order for the u's as functions of the x's, and the theory of such equations shows that these equations always possess a set of solutions (subject to certain conditions of regularity of the quantities a as functions of the x's). Hence m is less than or equal to $\frac{1}{2}n(n+1)$. We thus have, immediately, a classification of quadratic forms. A form is said to be of *rank* r when $r + n$ is the least value of m for which the form can be reduced to $\sum_{\lambda=1}^{m} du_\lambda^2$. Such a form may be interpreted geometrically as the square of the element of length of an n-fold manifold in ordinary (Euclidean) space of $n+r$ dimensions.

38. Differential parameters for forms of rank zero.

Let the form be of rank zero, then we have

$$\sum_{r,s=1}^{n} a_{rs}dx_r dx_s = \sum_{\lambda=1}^{n} du_\lambda^2,$$

and it is easy to prove that if one set of functions u is given, the most general possible set is given by performing a general orthogonal transformation and a translation on the set u. The problem in this particular case is therefore seen to separate into two parts:

(A) The determination of all invariants under a general transformation on the x's, subject to the condition that the u's are invariant.

(B) The selection from these of those functions which are still invariant when a translation and an orthogonal transformation are performed on the u's.

The first of these is equivalent to the determination of all the invariants of any number of functions of the set of variables x. If these functions are $f^{(1)}(x)$, $f^{(2)}(x)$, ..., the variables occurring in the invariants are

(a) derivatives of the f's,

(b) $x_1, \ldots, x_n$,

(c) $dx_1, \ldots, dx_n, d^2x_1, \ldots, d^2x_n, \ldots, d^px_n$.

We slightly extend the definition of an invariant by assuming that an invariant is a function I, that becomes $\Omega^\mu I$ under a general transformation of the group, where Ω is the Jacobian of the transformation, and μ is a number.

For an infinitesimal transformation Ω is easily seen to be

$$1 + \left(\sum_r \frac{\partial \xi_r}{\partial x_r}\right)\delta t,$$

and therefore the equation for I is

$$\frac{dI}{dt} = \mu\left(\sum_r \frac{\partial \xi_r}{\partial x_r}\right) I,$$

where $\frac{dI}{dt}\delta t$ denotes the effect of the infinitesimal transformation on I. The increments of the various variables in I may be obtained without difficulty. Let $f_{\alpha_1\alpha_2\ldots\alpha_n}$ denote $\frac{\partial^p f}{\partial x_1^{\alpha_1}\ldots\partial x_n^{\alpha_n}}$ where $\alpha_1+\alpha_2+\ldots\alpha_n = p$.

Then from the equation given above,

$$-\frac{d}{dt} f_{a_1 \dots a_n} = \sum_{i=1}^{n} (\xi_i)_{a_1 \dots a_n} \frac{\partial f}{\partial x_i} + \text{terms involving lower derivatives of the } \xi\text{'s.}$$

Also let $x^{(\lambda)}$ denote $d^\lambda x$, then it is at once clear that

$$\frac{d}{dt} x_j^{(\lambda)} = \left(\sum_{i=1}^{n} x_i^{(1)} \frac{\partial}{\partial x_i} \right)^\lambda \xi_j + \text{terms involving lower derivatives of the } \xi\text{'s.}$$

The expression

$$\left(\Sigma\, x_i^{(1)} \frac{\partial}{\partial x_i} \right)^\lambda$$

is supposed expanded by the multinomial theorem and then applied as an operator to ξ_j. Hence, if none of the increments of the variables in I contain derivatives higher than the pth of the ξ's, *i.e.* if I is an invariant of order less than $p+1$, the differential equations obtained by equating to zero the coefficients of these pth derivatives are

$$\sum_{k=1}^{n} \frac{\partial f^{(k)}}{\partial x_i} \frac{\partial I}{\partial f^{(k)}{}_{a_1 \dots a_n}} + \frac{p\,!}{a_1\,!\, a_2\,! \dots a_n\,!} (x_1^{(1)})^{a_1} \dots (x_n^{(1)})^{a_n} \frac{\partial I}{\partial x_i^{(p)}} = 0,$$

where $i = 1, 2, \dots, n$, and the a's take all positive integral and zero values subject to the condition that their sum is p. It is further assumed that there are n functions f.

A complete set of solutions of this system is readily seen to be given by $d^p f^{(1)}, d^p f^{(2)}, \dots, d^p f^{(n)}$, and it is clear that these expressions are invariants, and therefore they satisfy the equations obtained from lower derivatives of the ξ's. The remaining solutions of the complete set of equations arising from the coefficients of the ξ's and all their derivatives of orders up to p must therefore be of orders less than p. It therefore follows that the complete system of invariants is given by

$$d^\lambda f^{(k)}, \quad (\lambda = 2, 3, \dots p;\; k = 1, 2, \dots, n)$$

together with the solutions of the system of equations for invariants of the first order. A slight modification is here necessary because I occurs explicitly. It is easy to see that we have the solutions

$$df^{(1)}, df^{(2)}, \dots, df^{(n)},$$

and there remains yet one other solution, which is manifestly J, the Jacobian of the f's. Hence *a complete functionally independent system of invariants of n functions f of the variables $x_1, \dots, x_n$, involving derivatives and differentials up to the pth order, is given by*

$$d^\lambda f^{(k)}, \qquad (\lambda = 1, \dots, p;\; k = 1, \dots, n)$$

and J. For all except J, μ is zero, and for J μ is -1.

The most general invariant is therefore a function of the f's and the above absolute invariants, multiplied by some power of J.

39. Now let the n functions f be $u_1, u_2, \ldots, u_n$. The second part of our problem is to determine those invariants which still remain invariant when a general translation and a general orthogonal transformation are performed on the u's.

The variables entering are

(1) $u_1, u_2, \ldots, u_n$, and the total differentials of these variables;

(2) a number of arbitrary functions $\phi(u)$ and their derivatives, and this number we may as before take as not more than n;

(3) the Jacobian J.

The translation may be taken account of at once. It is equivalent to the condition that the variables u do not enter explicitly into the invariants.

The orthogonal transformation is a linear one on the variables u. We give the proof for two variables, though the method is quite general. Let $du^2 + dv^2 = dU^2 + dV^2$ where U, V are functions of u, v. Then if suffixes denote differentiations, this relation is equivalent to the three

$$U_1^2 + V_1^2 = 1, \quad U_1 U_2 + V_1 V_2 = 0, \quad U_2^2 + V_2^2 = 1.$$

Hence

$$J^2 \begin{pmatrix} U & V \\ u & v \end{pmatrix} \equiv (U_1^2 + V_1^2)(U_2^2 + V_2^2) - (U_1 U_2 + V_1 V_2)^2 = 1,$$

and therefore $J \begin{pmatrix} U & V \\ u & v \end{pmatrix}$ is not zero.

Also from the above three equations we deduce by differentiation with respect to u, v,

$$U_1 U_{11} + V_1 V_{11} = 0, \quad U_2 U_{11} + V_2 V_{11} + U_1 U_{12} + V_1 V_{12} = 0,$$
$$U_1 U_{12} + V_1 V_{12} = 0.$$

Hence $U_1 U_{11} + V_1 V_{11} = 0$, $U_2 U_{11} + V_2 V_{11} = 0$, and the determinant of the coefficients of these two linear equations in U_{11}, V_{11} is the Jacobian just mentioned, and is not zero. Hence U_{11}, V_{11} are both zero. Similarly all the other second derivatives of U and V are zero and hence U and V are linear functions of u and v. As a translation has already been taken account of, our transformation is homogeneous and linear on the variables u, v.

Now, returning to the case of n variables, let us consider a general homogeneous linear transformation on the u's; the variables $d^\lambda u$ are transformed by the same linear transformation. The first derivatives of a function ϕ are transformed by the contragredient transformation. In general, let A_m denote

$$\left\{ \sum_{i=1}^{n} U_i \frac{\partial}{\partial u_i} \right\}^m \phi,$$

where the U's are a set of auxiliary variables. Then A_m is an algebraic form of order m whose coefficients are the mth derivatives of ϕ, except for numerical multipliers. The transformations for these derivatives of ϕ are precisely those on the coefficients of the algebraic form A_m, when the U's are transformed by the contragredient transformation to the original linear transformation. Also J in the transformed variables is J in the original variables multiplied by the determinant of the transformation. If the transformation is orthogonal, it and its contragredient are the same; also its determinant is unity, hence J is an absolute invariant. We therefore immediately obtain the result:

The functionally independent set of invariants of orders up to and including the pth of the quadratic form $\sum_{i=1}^{n} du_i^2$ and n associated functions $\phi^{(1)}, \ldots, \phi^{(n)}$, are J, and the orthogonal algebraic invariants of the system of n-ary forms

$$A_\lambda^{(k)}, \qquad (k = 1, 2, \ldots, n;\ \lambda = 1, 2, \ldots, p)$$

$$\sum_{i=1}^{n} d^r u_i U_i. \qquad (r = 1, 2, \ldots, p)$$

The linear forms $\sum_{i=1}^{n} d^r u_i U_i$ may be omitted if account is taken of the invariants $d^r\phi^{(k)}$, and if the quadratic $\sum_{i=1}^{n} U_i^2$ is included we may state the result in terms of a general linear transformation, since an orthogonal linear transformation is a linear transformation which leaves the quadratic $\sum_{i=1}^{n} U^2$ invariant. The final result is:

The most general invariant is a function of the quantities $d^r\phi^{(k)}$, the general algebraic invariants of the forms A and of ΣU^2, multiplied by some power of J.

It is worthy of note that the algebraic forms A are the polar forms of the functions ϕ.

40. We now transform these algebraic forms from the variables U to variables X given by the scheme

$$U_i = \sum_{j=1}^{n} \frac{\partial u_i}{\partial x_j} X_j. \qquad (i = 1, 2, \ldots, n,)$$

The Jacobian of this transformation is J, and the discriminant of the quadratic form ΣU^2 changes from 1 to J^2. Hence the most general invariant is a function of the quantities $d^r\phi$, and of the general algebraic invariants of the A's and ΣU^2, expressed in the new variables.

Let $\sum_{i=1}^{n} U_i^2$ become $\sum_{r,s=1}^{n} a_{rs}X_rX_s$, then this is the fundamental quadratic form, and it will be shown that the coefficients of the forms A may be expressed in terms of the quantities a and their derivatives. We proceed to calculate these forms:

A_1 becomes $\sum_{i=1}^{n} X_i \frac{\partial\phi}{\partial x_i}$, and A_2 is readily seen to be $\sum_{r,s} \phi_{rs}X_rX_s$, where ϕ_{rs} denotes the covariant derivative of ϕ with respect to the fundamental quadratic form. In the general case

$$A_p = \sum_{i,j,\ldots,=1}^{n} \sum_{\lambda,\mu,\ldots,=1}^{n} \frac{\partial u_i}{\partial x_\lambda}\frac{\partial u_j}{\partial x_\mu} \cdots \frac{\partial^p \phi}{\partial u_i \partial u_j \ldots} X_\lambda X_\mu X_\nu \ldots,$$

where there are p letters $i, j, \ldots$, and also p letters $\lambda, \mu, \ldots$.

Let $\frac{p!}{\lambda!\mu!\ldots}\, {}_pB_{\lambda,\mu,\ldots}$ denote the general coefficient in the above expression for A_p. Then

$$\frac{\partial}{\partial x_\rho}\, {}_pB_{\lambda,\mu,\ldots} = {}_{p+1}B_{\rho,\lambda,\mu,\ldots} + \sum_{i,j,\ldots,=1}^{n} \frac{\partial^p\phi}{\partial u_i \partial u_j \ldots} \left\{ \Sigma \frac{\partial^2 u_i}{\partial x_\rho \partial x_\lambda} \frac{\partial u_j}{\partial x_\mu} \frac{\partial u_k}{\partial x_\nu} \cdots \right\},$$

where the last Σ denotes that there is a term corresponding to each of the letters $\lambda, \mu, \ldots$, and that these are added. Also

$${}_pB_{q,\mu,\nu,\ldots} = \sum_{i,j,\ldots,=1}^{n} \frac{\partial^p\phi}{\partial u_i \partial u_j \ldots} \frac{\partial u_i}{\partial x_q} \frac{\partial u_j}{\partial x_\mu} \cdots.$$

We solve the n equations obtained by giving q the values $1, \ldots, n$ for the quantities

$$\sum_{j,k,\ldots,=1}^{n} \frac{\partial^p\phi}{\partial u_i \partial u_j \ldots} \frac{\partial u_j}{\partial x_\mu} \frac{\partial u_k}{\partial x_\nu} \cdots$$

and substitute in the above equation. We thus obtain

$$\frac{\partial}{\partial x_\rho}\, {}_pB_{\lambda,\mu,\ldots} = {}_{p+1}B_{\rho,\lambda,\mu,\ldots} + \Sigma \left\{ \sum_{i=1}^{n} \sum_{q=1}^{n} {}_pB_{q,\mu,\nu,\ldots} M_q^i \frac{\partial^2 u_i}{\partial x_\rho \partial x_\lambda} \right\} \frac{1}{J},$$

where M_q^i is the cofactor of $\frac{\partial u_i}{\partial x_q}$ in J. This equation may be written

$${}_{p+1}B_{\rho,\lambda,\mu,\ldots} = \frac{\partial}{\partial x_\rho}\, {}_pB_{\lambda,\mu,\ldots} +$$

$$\Sigma \frac{1}{J} \begin{vmatrix} 0, & \frac{\partial^2 u_1}{\partial x_\rho \partial x_\lambda}, & \frac{\partial^2 u_2}{\partial x_\rho \partial x_\lambda}, & \ldots\ldots, & \frac{\partial^2 u_n}{\partial x_\rho \partial x_\lambda} \\ {}_pB_{1,\mu,\nu,\ldots}, & \frac{\partial u_1}{\partial x_1}, & \frac{\partial u_2}{\partial x_1}, & \ldots\ldots, & \frac{\partial u_n}{\partial x_1} \\ {}_pB_{2,\mu,\nu,\ldots}, & \frac{\partial u_1}{\partial x_2}, & \ldots\ldots & \ldots\ldots, & \frac{\partial u_n}{\partial x_2} \\ \vdots & \vdots & & & \vdots \\ {}_pB_{n,\mu,\nu,\ldots}, & \frac{\partial u_1}{\partial x_n}, & \ldots\ldots & \ldots\ldots, & \frac{\partial u_n}{\partial x_n} \end{vmatrix}.$$

We multiply the determinant in this equation by J, and note that $J^2 = \Delta$, the discriminant of the quadratic form, and thus obtain the result

$$
{}_{p+1}B_{\rho,\lambda,\mu,\ldots} = \frac{\partial}{\partial x_\rho}\,{}_pB_{\lambda,\mu,\ldots} + \\
\Sigma \frac{1}{\Delta} \begin{vmatrix} 0, & [\rho\lambda, 1], & [\rho\lambda, 2], & \ldots\ldots, & [\rho\lambda, n] \\ {}_pB_{1,\mu,\nu,\ldots}, & a_{11}, & a_{12}, & \ldots\ldots, & a_{1n} \\ {}_pB_{2,\mu,\nu,\ldots}, & a_{21}, & a_{22}, & \ldots\ldots, & a_{2n} \\ \vdots & \vdots & \vdots & & \vdots \\ {}_pB_{n,\mu,\nu,\ldots}, & a_{n1}, & a_{n2}, & \ldots\ldots, & a_{nn} \end{vmatrix}.
$$

The coefficients of the forms A_{p+1} are thus expressed in terms of those of the forms A_p, the a's and their derivatives, and we have shown that the coefficients of A_1 and A_2 may be expressed in terms of the derivatives of ϕ, the a's and their derivatives. It therefore follows by induction that the coefficients of A_p may be expressed in terms of these magnitudes. The process whereby ${}_{p+1}B_{\rho,\lambda,\mu,\ldots}$ is determined from ${}_pB_{\lambda,\mu,\ldots}$ is in fact that which has already been defined as covariant derivation, and thus the result is identified with that given by the earlier method.

The initial limitation that the quadratic form is of rank zero may now be removed, for it may be verified without difficulty that the invariants obtained are also invariants for a general quadratic form. When the form is of rank zero it has no other invariants than these, which are differential parameters. When the rank is greater than zero there are other invariants which do not involve any associated functions ϕ, namely the Gaussian invariants.

41. In the general case the quadratic is expressed as the sum of the square of m perfect differentials, where m is greater than n. Then it is clear that all the invariants of our quadratic must be included in those of the form $\sum_{\lambda=1}^{m} du_\lambda^2$; our form is in fact determined as an n dimensional manifold in Euclidean space of m dimensions, and it is obvious that among the invariants of the space must be included all the invariants of the manifold. The limitation is made by taking certain of the associated functions ϕ, equated to zero, to define the manifold. If we call them $x_{n+1}, \ldots, x_m$, we have the system of equations $x_{n+1} = 0, \ldots, dx_{n+1} = 0, \ldots$. By means of these equations the forms A for the parameters may be reduced so as to contain n instead of m variables X, and hence the covariant derivatives of a function ϕ are seen to lead to the parameters of a general quadratic form. The equations $x_{n+1}=0, \ldots, x_m=0$ lead to a set of algebraic forms whose coefficients are the generalisation of the fundamental magnitudes D, D', D'' * of a surface in space. These last are not invariants of the original quadratic form since they cease to be invariants if the n dimensional manifold is deformed in the m dimensional space. Among them, however, must be included the Gaussian invariants (those involving the a's and their derivatives only), but these last have not been determined by the method of Lie except in some of the simpler cases. An idea of the algebraic work necessary for a direct application of the Lie theory to the problem of the determination of Gaussian invariants may be gathered from a study of Forsyth's memoirs on the subject.

* See Bianchi-Lukat, *Differentialgeometrie*, p. 87, also Forsyth, *Mess. Math.* Vol. XXXII. (1903), pp. 68 *sqq*.

CHAPTER IV

MASCHKE'S SYMBOLIC METHOD

42. Explanation of the symbolic notation.

The remaining method for the determination of invariants, by means of a symbolic notation, is due to Maschke. Its leading principle will present no difficulty to those familiar with the ordinary symbolic notation of algebraic invariants.

The assumption is made that the quadratic form $\Sigma a_{ik}\, dx_i\, dx_k$ can be expressed symbolically $(df)^2$, where f is a symbolic function of the variables x. The derivatives of f, written $f_1, f_2, \ldots, f_n$, have no meaning when taken separately, but when their products are taken two at a time, $a_{ik} = f_i f_k$. Any expression involving the a's may thus be expressed in terms of the f's. If, however, the expression in the a's is of order higher than the first, we must avoid ambiguity by introducing equivalent symbols.

For example $f_1^2 f_2^2$ might be interpreted either as a_{12}^2 or $a_{11}a_{22}$, but if $\Sigma a_{ik} dx_i dx_k = (df)^2 = (d\phi)^2 = \ldots$, then $a_{12}^2 = f_1 f_2 \phi_1 \phi_2$, and $a_{11}a_{22} = f_1^2 \phi_2^2$, and there is no ambiguity. In general, enough equivalent symbols must be introduced to make the symbolic expression of the second order in the derivatives of each of the symbols f, ϕ, etc.

43. Invariantive constituent.

Now call the fundamental quadratic form A, and let $F^1, F^2, \ldots, F^n$ be any n invariant expressions of A, then we have

$$F^{i'} = F^i, \qquad (i = 1, \ldots, n)$$

where $F^{i'}$ is F^i for a new set of variables y. Hence

$$\sum_{k=1}^{n} \frac{\partial F^{i'}}{\partial y_k} dy_k = \sum_{k=1}^{n} \frac{\partial F^i}{\partial x_k} dx_k$$

and therefore

$$\left| \frac{\partial F^{i'}}{\partial y_k} \right| = r \left| \frac{\partial F^i}{\partial x_k} \right|,$$

where r is the Jacobian of the transformation, and $\left| \dfrac{\partial F^i}{\partial x_k} \right|$ denotes the

Jacobian of the F's with reference to the x's. Further, if $|a_{ik}|$ denotes the discriminant of A, $|a_{ik}'| = r^2 |a_{ik}|$, and therefore

$$|a_{ik}'|^{-\frac{1}{2}} \left|\frac{\partial F^{i'}}{\partial y_k}\right| = |a_{ik}|^{-\frac{1}{2}} \left|\frac{\partial F^i}{\partial x_k}\right|.$$

Hence $|a_{ik}|^{-\frac{1}{2}} \left|\frac{\partial F^i}{\partial x_k}\right|$ is an invariant of A.

This invariant may be denoted by (F). If it should be necessary to put in evidence the first one, two, three, etc. of the F's, they are written in their proper places and the last letter is assumed to run on. For example (b, c, a) denotes the invariant derived from $b, c, a^1, \ldots, a^{n-2}$, and similarly for other such expressions. (F) is called an *invariantive constituent* of A.

Now if A is written symbolically $df^2 = d\phi^2 = \ldots$, it is clear that f, ϕ, etc. are (symbolic) invariants. Hence if we form invariantive constituents of f, ϕ, etc. and any number of arbitrary functions of the independent variables, all products of these are formally invariant. These products will represent actual invariants whenever they are such that all the derivatives of the f's, ϕ's, etc. have an actual meaning. Further, an invariantive constituent may itself involve other invariantive constituents as elements, and thus higher derivatives than the first of the functions f, ϕ, etc. may occur. It is to be expected that some functions of the higher derivatives of f, for example, may be interpreted by means of the derivatives of magnitudes a_{ik}, and it is easy to see that any expression built up of invariantive constituents is an invariant if it can be interpreted in terms of actual quantities, for instance, the a's and their derivatives and the arbitrary functions introduced and their derivatives.

Now by actual differentiation

$$\frac{\partial a_{ik}}{\partial x_l} = f_i f_{kl} + f_k f_{il},$$

and there are similar expressions for $\frac{\partial a_{kl}}{\partial x_i}$, $\frac{\partial a_{il}}{\partial x_k}$.

These give at once

$$f_i f_{kl} = \tfrac{1}{2}\left[\frac{\partial a_{ik}}{\partial x_l} + \frac{\partial a_{il}}{\partial x_k} - \frac{\partial a_{kl}}{\partial x_i}\right] \equiv [kl, i].$$

Hence there is an interpretation for the symbolic product of type $f_i f_{kl}$.

Further, by differentiating again

$$f_{im} f_{kl} + f_i f_{klm} = [kl, i]_m,$$

and hence

$$f_{ir} f_{ks} - f_{kr} f_{is} = [ir, k]_s - [is, k]_r,$$

$$f_r f_{iks} - f_s f_{ikr} = [ik, r]_s - [ik, s]_r.$$

44. By means of the relations for second derivatives we have, for example,

$$\frac{1}{(n-1)!}((uf), f) = \Delta_2 u,$$

where $\Delta_2 u$ is Beltrami's second differential parameter, a result that may be verified by direct calculation. Some of the simpler invariants are*

$$(f)^2 = n!, \quad (uf)^2 = (n-1)!\,\Delta_1 u, \quad (uf)(vf) = (n-1)!\,\Delta(u, v).$$

In these examples f is used to denote a symbol by means of which the quadratic form is expressed, and u, v are actual functions of the variables. This notation will be generally followed, that is to say, f's distinguished by means of indices are the symbols of the quadratic form, the letters u, v, w, etc. are used to denote actual functions, and a, b, c, etc. are used in general formulae to denote quantities which may be interpreted either as symbols or as actual functions.

45. Symbolic identities.

Exactly as in the ordinary symbolism for algebraic invariants, there are many different symbolic expressions for a given invariant, due to the fact that equivalent symbols are used. For example, suppose an invariant to contain two equivalent symbols f and ϕ. This must have the same value as the expression got by interchanging f and ϕ. But regarded as an algebraic expression with f, ϕ as actual quantities, it may change sign when these two quantities are interchanged, and therefore, if it does so, it must vanish.

For many symbolic identities Maschke's paper quoted above should be consulted. We content ourselves with giving one or two illustrations.

We have first, since $(f)^2 = n!$, a constant,

$$(f)(f)_i = 0,$$

where the suffix denotes differentiation with respect to x_i. Now let $[f]$ denote (f) or any ordinary or covariant derivative of (f), and consider the product $f_1^1(uf)[f]$, where $f^1, f^2, \ldots, f^n$ are the equivalent symbols for f used in (f).

$$(uf) \equiv |a_{ik}|^{-\frac{1}{2}} \begin{vmatrix} u_1, & u_2, & \ldots, & u_n \\ f_1^2, & f_2^2, & \ldots, & f_n^2 \\ \cdots & \cdots & \cdots & \cdots \\ \cdots & \cdots & \cdots & \cdots \\ f_1^n, & f_2^n, & \ldots, & f_n^n \end{vmatrix} = |a_{ik}|^{-\frac{1}{2}} \{uf\}, \text{ say,}$$

* See Maschke, *Trans. Amer. Math. Soc.* Vol. IV. p. 441 *sqq.* (1903).

and
$$f_1^1\{uf\}[f]=\begin{vmatrix} f_1u_1, & u_2, & \ldots, & u_n \\ f_1^1f_1^2, & f_2^2, & \ldots, & f_n^2 \\ \vdots & \vdots & & \vdots \\ f_1^1f_1^n, & f_2^n, & \ldots, & f_n^n \end{vmatrix}[f].$$

Now if $f_1^1f_1^k[f]$ has an actual interpretation so far as f^1 and f^k are concerned; also it changes sign if these two equivalent symbols are interchanged. Hence it vanishes, and therefore

$$f_1^1\{uf\}[f]=f_1^1u_1\begin{vmatrix} f_2^2 & \ldots, & f_n^2 \\ \vdots & & \vdots \\ f_2^n, & \ldots, & f_n^n \end{vmatrix}[f],$$

and this expression is easily seen to be equal to

$$u_1\begin{vmatrix} f_1^1, & f_2^1, & \ldots, & f_n^1 \\ 0, & f_2^2, & \ldots, & f \\ 0, & \vdots & & \vdots \\ \vdots & & & \\ 0, & f_2^n, & \ldots, & f_n^n \end{vmatrix}[f].$$

If in the above determinant we interchange the equivalent symbols f^1 and f^p, and let p take all values from 2 to n, we have n different forms for $f_1^1\{uf\}[f]$. If these are added, the determinants combine into $\{f\}$, and it follows finally that

$$f_1^1(uf)[f]=\frac{1}{n}u_1(f)[f].$$

Similarly
$$f_i^1(uf)[f]=\frac{1}{n}u_i(f)[f],$$

or changing the notation, we have

$$f_i(ua)[fa]=\frac{1}{n}u_i(fa)[fa],$$

where a denotes a symbol.

For example, if $[fa]$ denotes (fa) or $(fa)_i$ we have the two results

$$f_k(fa)(ua)=(n-1)!\,u_k,$$
$$f_k(fa)_i(ua)=0.$$

If $v^2, v^3, \ldots, v^n$ are arbitrary functions we may readily deduce from these two results

$$(fa)(fv)(ua)=(n-1)!\,(uv),$$
$$(fa)_i(fv)(ua)=0.$$

By a method precisely similar to that given above we may prove that $f_1^1f_2^2(uvf)[f]$ is equal to

$$\frac{1}{n(n-1)}(u_iv_k-u_kv_i)(f)[f],$$

and hence, for example,

$$f_1\phi_k\,(uva)\,(f\phi a) = (n-2)\,!\,(u_i v_k - u_k v_i)$$

$$f_i\phi_k\,(f\phi a) = 0.$$

By differentiation of these results we have *e.g.* if x is any function of the independent variables,

$$[f_{ik}\,(xa) + f_i\,(xa)_k]\,[fa] = \frac{1}{n}\,[x_{ik}\,(fa) + x_i\,(fa)_k]\,[fa],$$

and it easily follows that x_{ik} is equal to

$$\frac{1}{(n-1)\,!}\{f_i\,(fa)\,(xa)_k + f_{ik}\,(fa)\,(xa)\}.$$

Again, it is easy to show that $f_k\,(ua)_i\,(fa)$ is equal to

$$f_i\,(ua)_k\,(fa),$$

and assuming f and a to be symbols and u a function of the independent variables, we have the covariant

$$\sum_{ik}\,(fa)\,f_i\,(ua)_k\,dx_i dx_k,$$

whose coefficients can be shown to be equal to the second covariant derivatives of the function u. This covariant may be written

$$(fa)\,f_x\,(ua)_x.$$

46. For further work we require the higher covariant derivatives of a given function of the variables. Following Maschke's notation (which differs from that used in Chapter II) we use x to denote any function not involving derivatives, and $x^{(ik\cdots)}$ for a covariant derivative of x, suffixes still being used to denote ordinary derivatives.

We have

$$x^{(\lambda)} = x_\lambda, \quad x^{(ik)} = x_{ik} - \epsilon f_{ik}\,(fa)\,(xa),$$

where $\epsilon \equiv \frac{1}{(n-1)\,!}$, and f and a are equivalent symbols for the fundamental form. It follows by means of the expression already calculated for x_{ik} that $x^{(ik)} = \epsilon f_i\,(xa)_k\,(fa)$. The higher covariant derivatives can now be calculated as soon as we know the covariant derivative of an invariantive constituent, for we know that to differentiate a product covariantively we apply the same rule as in ordinary differentiation, *e.g.*

$$\{x_i y_k\}^{(\lambda)} = x_i y^{(k\lambda)} + y_k x^{(i\lambda)}.$$

It can be proved that

$$(f)^{(\lambda)} = (f)_\lambda,$$

and the higher derivatives may be calculated at once; for example we deduce from the above expression for $x^{(ik)}$ that

$$x^{(ik\lambda)} = \epsilon \{ f^{(i\lambda)} (xa)_k (fa) + f_i (xa)^{(k\lambda)} (fa) + f_i (xa)_k (fa)_\lambda \}.$$

Also $f^{(i\lambda)} = \epsilon \phi_i (fa)_\lambda (\phi a)$, where ϕ is a symbol, and therefore

$$f_k f^{(i\lambda)} = \epsilon f_k \phi_i (fa)_\lambda (\phi a).$$

We have proved previously that

$$f_k (fa)_\lambda (\phi a) = 0;$$

hence $f_k f^{(i\lambda)} = 0$, that is to say, the product of a first and second covariant derivative of a symbol vanishes, and hence also

$$(fu) f^{(ik)} = 0.$$

Therefore

$$x^{(ik\lambda)} = \epsilon \{ f_i (fa) (xa)^{(k\lambda)} + f_i (xa)_k (fa)_\lambda \},$$

and $(xa)^{(k\lambda)}$ can be calculated from the formula for second derivatives.

47. Symbolic expressions for the forms G.

The expression of the quadrilinear form

$$G_4 \equiv \sum_{i, k, r, s} (ikrs)\, d^{(1)}x_i d^{(2)}x_k d^{(3)}x_r d^{(4)}x_s$$

in symbols is next required.

From the definition of $(ikrs)$ it is not difficult to deduce that

$$\begin{aligned} (ikrs) &= f_{ir} f_{ks} - f_{is} f_{kr} + \epsilon \{ f_{is}\phi_{kr} - f_{ir}\phi_{ks} \} (\phi a)(fa) \\ &= f_{ir} \{ f_{ks} - \epsilon \phi_{ks} (\phi a)(fa) \} - f_{is} \{ f_{kr} - \epsilon \phi_{kr} (\phi a)(fa) \} \\ &= f_{ir} f^{(ks)} - f_{is} f^{(kr)}; \end{aligned}$$

this may also be written

$$f^{(ir)} f^{(ks)} - f^{(is)} f^{(kr)}.$$

Many different symbolic expressions may now be deduced for $(ikrs)$, for example

$$(ikrs) = \epsilon f_i \phi_k \{ (fa)_r (\phi a)_s - (fa)_s (\phi a)_r \},$$

and it is not hard to prove that

$$f_i \phi_k (uva)(f\phi a) = (n-2)!\, (u_i v_k - u_k v_i).$$

Using this, we have

$$(n-1)!\,(n-2)!\,(ikrs) = ((fa)(\phi a) b)(f'\phi' b) f_i \phi_k f_r' \phi_s',$$

and therefore

$$(n-1)!\,(n-2)!\,G_4 = ((fa)(\phi a) b)(f'\phi' b) f_{x^1} \phi_{x^2} f_{x^3}' \phi_{x^4}'.$$

G_5 may now be calculated, and

$$\begin{aligned} &(n-1)!\,(n-1)!\,(n-2)!\,G_5 \\ &\quad = (((fu)(\phi a) b) c)(f'\phi' b)(f'' c) f_{x^1} \phi_{x^2} f_{x^3}' \phi_{x^4}' f_{x^5}'', \end{aligned}$$

and there are similar symbolic expressions for G_6, G_7, etc. These symbolic expressions for the G's put in evidence at once the fact that they are invariants.

48. For further work on this part of the subject Maschke's papers should be consulted. We give, however, one interesting application of the theory, taken from one of his papers*.

Let f, ϕ, ψ, etc. denote symbols, and let any two dimensional manifold be determined in the n-dimensional space (not necessarily Euclidean) for which ds^2 is equal to the fundamental quadratic form. This manifold is given by $n-2$ equations

$$U_1,\ U_2,\ \ldots,\ U_{n-2}=0.$$

Maschke has proved that for this manifold

$$K=\frac{2\left(\left(f\psi U\right)\left(\phi\psi U\right)U\right)\left(f\phi U\right)}{\left(f'\phi' U\right)^2\left(f''\phi'' U\right)^2}.$$

If $n=2$ the U's disappear and this expression becomes

$$K=\frac{1}{2\left(EG-F^2\right)}\left(2\frac{\partial^2 F}{\partial u\partial v}-\frac{\partial^2 E}{\partial v^2}-\frac{\partial^2 G}{\partial u^2}\right)+\ldots,$$

where $$ds^2=Edu^2+2Fdudv+Gdv^2.$$

This is the well-known Gaussian expression for K.

Next let $n=3$ and let the space be Euclidean. Then we may take

$$ds^2=dx^2+dy^2+dz^2,$$

and there is one expression U. Take the equation of the surface to be

$$F(x,\ y,\ z)=0,$$

then U is F. If we substitute in the above general expression for K we get

$$K=-\frac{1}{\left(\frac{\partial F}{\partial x}\right)^2+\left(\frac{\partial F}{\partial y}\right)^2+\left(\frac{\partial F}{\partial z}\right)^2}\begin{vmatrix}\frac{\partial^2 F}{\partial x^2}, & \frac{\partial^2 F}{\partial x\partial y}, & \frac{\partial^2 F}{\partial x\partial z}, & \frac{\partial F}{\partial x}\\ \frac{\partial^2 F}{\partial x\partial y}, & \frac{\partial^2 F}{\partial y^2}, & \frac{\partial^2 F}{\partial y\partial z}, & \frac{\partial F}{\partial y}\\ \frac{\partial^2 F}{\partial x\partial z}, & \frac{\partial^2 F}{\partial y\partial z}, & \frac{\partial^2 F}{\partial z^2}, & \frac{\partial F}{\partial z}\\ \frac{\partial F}{\partial x}, & \frac{\partial F}{\partial y}, & \frac{\partial F}{\partial z}, & 0\end{vmatrix}$$

The two expressions for K given above have no apparent connection. They are, however, two interpretations of one and the same general formula.

* Maschke, *Trans. Amer. Math. Soc.* (1906), Vol. VI. p. 93.

CHAPTER V

APPLICATIONS

49. We now consider some applications of the theory of differential invariants. Suppose first that the quadratic form is interpreted as the square of the element of length in a certain n-dimensional manifold. The invariants, though expressed analytically, are intrinsically connected with the manifold, apart from any frame of reference. They are differential, and involve only one set of the independent variables. They thus give the intrinsic character of the manifold at, and infinitely near, a particular point of it. If we take account of differential parameters, they express quantities intrinsically connected with the section of the fundamental manifold by another manifold.

But we have a set of quantities, defined geometrically, of just this type, that is to say they depend only on the manifold at or near some particular point and are intrinsically connected with it.

Let us consider more in detail the particular case of a surface in Euclidean space of three dimensions. This is not intrinsically determinate if only one quadratic form, that for ds^2, is given. For example, all developable surfaces have the same quadratic form for ds^2 as that of a plane. The catenoid (the surface of revolution obtained by revolving a catenary about its directrix) and the regular helicoid (the surface swept out by a line which is parallel to a fixed plane, intersects a fixed line perpendicular to the plane, and rotates uniformly about the fixed line as its point of intersection moves uniformly along that line) are two surfaces with the same quadratic form for ds^2. Two surfaces which have the same quadratic form for ds^2 are said to be *applicable* to each other. If we regard a surface as made of some perfectly flexible inextensible material, then it is clear that the surfaces into which it may be deformed are the surfaces applicable to it.

There is a second quadratic form which we may associate with a given surface, namely that for ds^2/ρ, where $1/\rho$ is the curvature of the normal section by a plane which meets the surface in the direction du/dv. The coefficients of the second form cannot be chosen arbitrarily if the first is given, though they involve a certain amount of arbitrariness.

If both these forms are given, the surface is intrinsically determinate. It follows that the geometrical magnitudes at a particular point on the surface are the differential invariants of *two* quadratic forms. Similarly, the geometrical quantities connected with any curve $U = 0$ on the surface may be expressed in terms of differential parameters that involve two quadratic forms. Thus the principal radii of curvature of the surface, the normal, geodesic, and principal curvatures and the torsion of a given curve on the surface are differential parameters of this type. Among all these geometrical quantities, there are some that are the same for all surfaces applicable to each other. Such are, for example, the total curvature $1/R_1R_2$ of the surface, the angle of intersection of two curves on the surface, the geodesic curvature of any curve on the surface, and many others. These, it is clear, do not depend on the second quadratic form, and they are therefore invariants and parameters of a single quadratic form, that for ds^2.

It follows, conversely, that any invariant or parameter of the form for ds^2 represents some geometrical magnitude associated with the surface, and the magnitude is the same for all surfaces applicable to that given. On this account, such invariants are called deformation invariants.

50. Geometrical interpretation of invariants.

In order to apply the invariant theory, we have now to interpret our invariants in terms of geometrical magnitudes, and also to interpret geometrical magnitudes in terms of invariants. It is perhaps worth while to point out the advantages gained by the use of invariant theory.

In the first place we are able to apply in the simplest possible way the methods of analysis to geometrical problems, for we express all our data analytically, and yet avoid extraneous properties which arise through the relations of our surface to a particular coordinate frame chosen. Again, when we express a given quantity as far as possible by means of invariants, it may happen that its expression only involves invariants, and thus its invariance becomes intuitive. Also, if we know that a given quantity is invariant, we can often determine its invariantive form for some simple choice of coordinates, and then we are able to write down its expression for any coordinates whatever. An example of this has already been given in the case of Laplace's equation

$$\frac{\partial^2 V}{\partial x^2} + \frac{\partial^2 V}{\partial y^2} + \frac{\partial^2 V}{\partial z^2} = 0.$$

In this case we are dealing with Euclidean space of three dimensions for which $ds^2 = dx^2 + dy^2 + dz^2$, and the equation, in the notation of the Absolute Differential Calculus, is $\sum_{r,s} a^{(rs)} V_{rs} = 0$, for this particular quadratic form. It follows at once that

$$\frac{\partial^2 V}{\partial x^2} + \frac{\partial^2 V}{\partial y^2} + \frac{\partial^2 V}{\partial z^2},$$

is an invariant of a general quadratic form in three variables, and its general expression is $\Sigma a^{(rs)} V_{rs}$ for any such quadratic form.

Another example is given by Darboux (*Theory of Surfaces*, Vol. III. p. 201). Let the quadratic form be

$$ds^2 = A^2 du^2 + C^2 dv^2,$$

then the geodesic curvature of the curves $v = \text{const.}$ is

$$-\frac{1}{AC}\frac{\partial A}{\partial v}.$$

If
$$\Delta\phi = \Sigma a^{(rs)} \phi_r \phi_s, \quad \Delta(\phi, \psi) = \Sigma a^{(rs)} \phi_r \psi_s,$$
$$\Delta_2 \phi = \Sigma a^{(rs)} \phi_{rs},$$

then these three quantities are obviously invariants, and

$$\Delta v = \frac{1}{C^2}, \quad \Delta(v, \Delta v) = -\frac{2}{C^5}\frac{\partial C}{\partial v},$$

$$\Delta_2 v = \frac{1}{AC}\frac{\partial}{\partial v}\left(\frac{A}{C}\right) = -\frac{1}{C^3}\frac{\partial C}{\partial v} - \frac{1}{C\rho}.$$

Hence
$$\frac{1}{\rho} = -\frac{\Delta_2 v}{\sqrt{\Delta v}} + \frac{1}{2}\frac{\Delta(v, \Delta v)}{(\Delta v)^{\frac{3}{2}}}.$$

It follows that $1/\rho$ is an invariant, since the Δ's are all invariants, and further, the geodesic curvature of any curve $\phi(u, v) = \text{const.}$, on a surface for which the quadratic form ds^2 is perfectly general, is

$$-\frac{\Delta_2 \phi}{\sqrt{\Delta\phi}} + \frac{1}{2}\frac{\Delta(\phi, \Delta\phi)}{(\Delta\phi)^{\frac{3}{2}}}.$$

This example is a good illustration of the advantages derived from the invariant theory. We start with

$$-\frac{1}{AC}\frac{\partial A}{\partial v},$$

and express it as far as possible in terms of invariants. It happens that it is entirely expressed by means of invariants, and hence it is itself an invariant. Further, when it is thus expressed in terms of invariants only, its general value may be at once written down.

We also note that the differential equation for all families of geodesics on the given surface is

$$2\Delta_2\phi \, . \, \Delta\phi = \Delta(\phi, \Delta\phi),$$

and this may be turned into a differential equation for the geodesics themselves by writing

$$\frac{dv}{du} = -\frac{\partial\phi}{\partial u} \Big/ \frac{\partial\phi}{\partial v},$$

with a corresponding expression for the second derivative

$$\frac{d^2v}{du^2}\text{*}.$$

51. Another application of the differential invariant theory can be made by means of the theory of algebraic invariants. All our differential invariants and parameters are invariants of algebraic forms. By means of the algebraic theory we can determine syzygies or algebraic relations connecting these invariants, and such syzygies, when expressed in terms of the geometrical magnitudes of the surface, lead to algebraic relations among these apparently independent quantities. Also the coefficients of the various forms are algebraically independent, and thus all such relations are given by syzygies. A surface is intrinsically determinate if two quadratic forms are given, and hence all its geometrical magnitudes are invariants of two quadratic forms. We are thus able, by means of the invariant theory, to determine which among these magnitudes are independent, and, by means of syzygies, to determine all the relations connecting those that are not independent. For illustrations of this part of the subject the reader should consult the latter part of Forsyth's memoir †.

52. The case of a surface in ordinary space—the quadratic form in two variables.

Confining ourselves to a quadratic form in two variables, we suppose that there are associated with the form two functions $\phi(u, v)$, $\psi(u, v)$. In this case there is only one Riemann symbol, the quantity G. We

* For other examples the reader should consult Wright, *Bull. Amer. Math. Soc.* Vol. XII. (1906), p. 379.

† *Phil. Trans.* (1903), Ser. A, vol. 201, pp. 369 *sqq.*

may therefore take instead of the series of forms G_4, G_5, ..., the quantity K, which is an absolute invariant, and the covariant derived forms of K. The set of algebraic forms is now

(i) the fundamental quadratic,

(ii) ϕ, ψ, K, and their successive covariant derived forms.

All the deformation invariants and parameters of a surface are therefore given by the algebraic invariants of these forms. If the quadratic form is $Edu^2 + 2Fdudv + Gdv^2$, we have the invariants

$$\Delta\phi \equiv \Sigma a^{(rs)}\phi_r\phi_s \equiv \frac{1}{H^2}\left\{E\left|\frac{\partial\phi}{\partial v}\right|^2 - 2F\frac{\partial\phi}{\partial u}\frac{\partial\phi}{\partial v} + G\left|\frac{\partial\phi}{\partial u}\right|^2\right\},$$

$$\Delta(\phi, \psi) \equiv \Sigma a^{(rs)}\phi_r\psi_s \equiv \frac{1}{H^2}\left\{E\frac{\partial\phi}{\partial v}\frac{\partial\psi}{\partial v} - F\left(\frac{\partial\phi}{\partial u}\frac{\partial\psi}{\partial v} + \frac{\partial\phi}{\partial v}\frac{\partial\psi}{\partial u}\right) + G\frac{\partial\phi}{\partial u}\frac{\partial\psi}{\partial u}\right\},$$

$$\Theta(\phi, \psi) \equiv \Sigma\epsilon^{(rs)}\phi_r\psi_s \equiv \frac{1}{H}\left\{\frac{\partial\phi}{\partial u}\frac{\partial\psi}{\partial v} - \frac{\partial\psi}{\partial u}\frac{\partial\phi}{\partial v}\right\},$$

where H is written for $\sqrt{EG - F^2}$. These are obtained from the quadratic and two linear forms; they are not independent, for

$$\Delta^2(\phi, \psi) + \Theta^2(\phi, \psi) = \Delta\phi \,.\, \Delta\psi,$$

and we thus have an example of a syzygy between invariants. This has a simple geometrical interpretation, for if the curves ϕ and ψ cut at an angle α, it may be easily proved that

$$\cos\alpha = \frac{\Delta(\phi, \psi)}{\sqrt{\Delta\phi \,.\, \Delta\psi}}, \quad \sin\alpha = \frac{\Theta(\phi, \psi)}{\sqrt{\Delta\phi \,.\, \Delta\psi}},$$

and the interpretation is therefore $\cos^2\alpha + \sin^2\alpha = 1$.

The invariants are all interpreted geometrically when the geometrical interpretation of $\Delta\phi$ is obtained. Let dn denote the perpendicular distance between the curves ϕ and $\phi + d\phi$ at the point (u, v), then

$$\frac{d\phi}{dn} = \frac{\partial\phi}{\partial u}\frac{du}{dn} + \frac{\partial\phi}{\partial v}\frac{dv}{dn}.$$

Also, if ds denote an infinitesimal arc of the curve $\phi = \text{const.}$,

$$\frac{\partial\phi}{\partial u}\frac{du}{ds} + \frac{\partial\phi}{\partial v}\frac{dv}{ds} = 0,$$

$$E\frac{du}{ds}\frac{du}{dn} + F\left(\frac{du}{ds}\frac{dv}{dn} + \frac{dv}{ds}\frac{du}{dn}\right) + G\frac{dv}{ds}\frac{dv}{dn} = 0,$$

since the directions ds and dn are perpendicular to each other. Hence

$$\frac{du}{dn} : \frac{dv}{dn} = G\frac{\partial\phi}{du} - F\frac{\partial\phi}{dv} : E\frac{\partial\phi}{dv} - F\frac{\partial\phi}{du},$$

and since $$E\left(\frac{du}{dn}\right)^2 + 2F\frac{du}{dn}\frac{dv}{dn} + G\left(\frac{dv}{dn}\right)^2 = 1,$$

we see that each of these ratios is equal to $1/H^2\sqrt{\Delta\phi}$, and therefore $d\phi/dn = \sqrt{\Delta\phi}$. We thus have the geometrical interpretation of $\Delta\phi$, namely, it is the square of $d\phi/dn$.

53. An important parameter of the second order is that known as Beltrami's second differential parameter

$$\Delta_2\phi \equiv \Sigma a^{(rs)}\phi_{rs} \equiv \frac{1}{H}\frac{\partial}{\partial u}\left\{\frac{1}{H}\left(G\frac{\partial\phi}{\partial u} - F\frac{\partial\phi}{\partial v}\right)\right\} + \frac{1}{H}\frac{\partial}{\partial v}\left\{\frac{1}{H}\left(E\frac{\partial\phi}{\partial v} - F\frac{\partial\phi}{\partial u}\right)\right\}.$$

Beltrami's method of obtaining it is by an application of Green's Theorem. In fact it is easy to prove that if ϕ and ψ are two functions regular inside a closed contour on the surface, and if $d\sigma$ denotes an element of area of the contour, ds an arc of the boundary,

$$\iint\Delta(\phi, \psi)\, d\sigma = -\int\psi\frac{\partial\phi}{\partial n}\, ds - \iint\psi\,\Delta_2\phi\, d\sigma,$$

and it follows, since all the other terms of the equation are invariants, and since the contour is arbitrary, that $\Delta_2\phi$ is an invariant.

54. All the parameters may be expressed in terms of three of them, and their derivatives.

The invariants Δ, Θ, Δ_2 once obtained, we can calculate from them many others, for example

$$\Delta\Delta\phi,\quad \Delta(\phi, \Delta\phi),\quad \Delta_2(\Delta\phi),\quad \Delta(\Delta\phi, \Delta\psi), \text{ etc.}$$

We shall prove that all invariants can be obtained by algebraic combinations of these, a result due to Beltrami. Suppose that the quadratic form has been transformed so that ϕ, ψ are the parametric curves, then the coefficients may be calculated without trouble and we have

$$ds^2 = \frac{1}{\Theta^2(\phi, \psi)}\{\Delta\psi d\phi^2 - 2\Delta(\phi, \psi)\, d\phi d\psi + \Delta\phi d\psi^2\}.$$

Now let I be any invariant involving the two functions ϕ, ψ; if ϕ, ψ are taken as independent variables, the quadratic form becomes that just given, and I is a function of the invariants Δ and their derivatives with respect to ϕ and ψ. But if λ is any function whatever, it follows from the definition of Θ that

$$\frac{\partial\lambda}{\partial\phi} = \Theta(\lambda, \psi)/\Theta(\phi, \psi).$$

Hence all the derivatives may be expressed by means of the Θ's, and the result follows. If the invariant contains only one function ϕ, we may take ϕ, and either $\Delta\phi$ or $\Delta_2\phi$, as the two independent variables of the form. It follows, in this case, that all the invariants may be obtained by repeated application of the operations Δ, Δ_2 to the single function ϕ.

Although we thus have a method whereby all invariants can be obtained, the result is not complete, for we have no clue as to the independence of the invariants, and frequently they are not thus expressed in their simplest forms. We know from the algebraic theory that there are three independent invariants of the second order involving one function ϕ. These must be

$$\Delta\Delta\phi, \quad \Delta(\phi, \Delta\phi), \quad \Delta_2\phi.$$

There are four of the third order, and these are included in

$$\Delta\Delta\Delta\phi, \quad \Delta(\phi, \Delta\Delta\phi), \quad \Delta(\Delta\phi, \Delta\Delta\phi), \quad \Delta(\Delta_2\phi), \quad \Delta(\phi, \Delta_2\phi), \text{ etc.},$$

but we have no immediate method of showing what are the relations that connect the last set.

55. Another method for deriving invariants of order $r+1$ from those of order r is the following: Let I be any invariant, and let ds_1 denote an arc of the curve $\phi = \text{const.}$ We have

$$\frac{dI}{ds_1} = \frac{\partial I}{\partial u}\frac{du}{ds_1} + \frac{\partial I}{\partial v}\frac{dv}{ds_1}, \qquad \frac{\partial \phi}{\partial u}\frac{du}{ds_1} + \frac{\partial \phi}{\partial v}\frac{dv}{ds_1} = 0,$$

$$E\left(\frac{du}{ds_1}\right)^2 + 2F\left(\frac{du}{ds_1}\right)\left(\frac{dv}{ds_1}\right) + G\left(\frac{dv}{ds_1}\right)^2 = 1.$$

Hence dI/ds_1, which is obviously an invariant, and is of order one higher than I, is $\Theta(I, \phi)/\sqrt{\Delta\phi}$. Similarly $dI/ds_2 = \Theta(I, \psi)/\sqrt{\Delta\psi}$, where ds_2 is an arc of the curve $\psi = \text{const.}$, and any function of these two quantities is an invariant of order $r+1$, where I is of order r. In particular, suppose that ϕ, ψ are such that $\Delta(\phi, \psi) = 0$, then the curves $\phi = \text{const.}$, $\psi = \text{const.}$ cut at right angles. An invariant of order $r+1$ is

$$\left(\frac{dI}{ds_1}\right)^2 + \left(\frac{dI}{ds_2}\right)^2,$$

and this is readily seen to be ΔI; it thus does not involve ϕ, ψ explicitly. For example, the Gaussian invariant of the third order may be written either ΔK or $(dK/ds_1)^2 + (dK/ds_2)^2$, where ds_1, ds_2 are any two infinitesimal arcs through u, v cutting at right angles.

56. Geometrical properties expressed by the vanishing of invariants.

We next consider the geometrical properties involved in the vanishing of certain invariants. If $\Theta(\phi, \psi) = 0$, from what we have already seen, ϕ and ψ cut at an angle zero; therefore they are coincident, that is to say, ψ is a function of ϕ. If $\Delta(\phi, \psi) = 0$ the curves ϕ and ψ cut at right angles. Now $\Delta\phi$ is $\Delta(\phi, \phi)$ and hence, if $\Delta\phi = 0$, the curve $\phi = \text{const.}$ is at right angles to itself. This curve (of course imaginary) must therefore be such that the tangent at any point meets the circle at infinity. Such curves are analogues of the straight lines $y + ix = k$ in a plane. Along such a curve

$$\frac{\partial\phi}{\partial u} du + \frac{\partial\phi}{\partial v} dv = 0,$$

and it therefore follows that $Edu^2 + 2Fdudv + Gdv^2 = 0$, that is to say $ds^2 = 0$. The curves are hence such that the distance between any two points on one of them, measured along that curve, is zero.

Consider the more general case of $\Delta\phi = f(\phi)$. Let the curves ψ be chosen at right angles to the curves ϕ, then $\Delta(\phi, \psi) = 0$, and therefore

$$ds^2 = \frac{d\phi^2}{\Delta\phi} + \frac{d\psi^2}{\Delta\psi},$$

or if $du = d\phi/\sqrt{f(\phi)}$, $ds^2 = du^2 + C^2 d\psi^2$, where C is a function of u and ψ. The curves $u = \text{const.}$ and the curves $\phi = \text{const.}$ are obviously the same, since u is a function of ϕ. Now if $1/\rho$ is the geodesic curvature of the curves $\psi = \text{const.}$, and

$$ds^2 = A^2 du^2 + C^2 d\psi^2, \quad \frac{1}{\rho} = -\frac{1}{AC}\frac{\partial A}{\partial v}.$$

Hence, in our case, $1/\rho$ is zero. The ψ curves are therefore geodesics, and the u's cut them orthogonally. Hence, if $\Delta\phi = f(\phi)$, the curves $\phi = \text{const.}$ are the orthogonal trajectories of a family of geodesics.

57. In order to develope the properties of $\Delta_2\phi$, we consider the following problem. It is proposed to express the fundamental quadratic in the form

$$P(x)\, dp^2 + 2Q(x)\, dpdq + R(x)\, dq^2,$$

where P, Q, R are given functions of the single argument x, and x, p, q are functions of u, v to be determined.

We have $\Delta p = R/(PR - Q^2) = S$, say, then S is a known function of x. Also

$$\Delta(p, \Delta p) = S' \frac{R\dfrac{\partial x}{\partial p} - Q\dfrac{\partial x}{\partial q}}{PR - Q^2}, \quad \Theta(p, \Delta p) = S' \frac{\dfrac{\partial x}{\partial q}}{\sqrt{PR - Q^2}},$$

and

$$\Delta_2 p = \frac{1}{\sqrt{PR - Q^2}} \left\{ \frac{\partial}{\partial p} \frac{R}{\sqrt{PR - Q^2}} - \frac{\partial}{\partial q} \frac{Q}{\sqrt{PR - Q^2}} \right\}.$$

From these equations, by elimination of x, $\partial x/\partial p$, $\partial x/\partial q$, there is obtained a relation between the invariants of the first and second orders of p. This relation, being invariantive, may be written in terms of the original form, and we thus have an equation of the second order for p as a function of u, v. Taking any solution of this equation, we obtain without difficulty an equation of the first order for q, and then x may be obtained by elimination.

Consider, for example, the case in which $Q = 0$, $R = P = x$. We now have for the quadratic form, $x(dp^2 + dq^2)$, and

$$\Delta p = \frac{1}{x}, \quad \Delta(p, \Delta p) = -\frac{1}{x^3}\frac{\partial x}{\partial p}, \quad \Theta(p, \Delta p) = \frac{1}{x^3}\frac{\partial x}{\partial q},$$

$$\Delta_2 p = 0.$$

The second order differential equation is thus $\Delta_2 p = 0$.

Conversely, let p be any solution of $\Delta_2 p = 0$, and let a function q be chosen so that $\Delta(p, q) = 0$. We have

$$\frac{\partial}{\partial u}\left\{\frac{1}{H}\left(G\frac{\partial p}{\partial u} - F\frac{\partial p}{\partial v}\right)\right\} + \frac{\partial}{\partial v}\left\{\frac{1}{H}\left(E\frac{\partial p}{\partial v} - F\frac{\partial p}{\partial u}\right)\right\} = 0,$$

$$E\frac{\partial p}{\partial v}\frac{\partial q}{\partial v} - F\left(\frac{\partial p}{\partial u}\frac{\partial q}{\partial v} + \frac{\partial p}{\partial v}\frac{\partial q}{\partial u}\right) + G\frac{\partial p}{\partial u}\frac{\partial q}{\partial u} = 0.$$

From the first of these equations it follows that

$$\frac{1}{H}\left(G\frac{\partial p}{\partial u} - F\frac{\partial p}{\partial v}\right) = \frac{\partial f}{\partial v}, \quad \frac{1}{H}\left(E\frac{\partial p}{\partial v} - F\frac{\partial p}{\partial u}\right) = -\frac{\partial f}{\partial u},$$

where f is a certain function of u and v, and the second equation becomes

$$\frac{\partial f}{\partial u}\frac{\partial q}{\partial v} - \frac{\partial f}{\partial v}\frac{\partial q}{\partial u} = 0.$$

Hence q is a function of f, and without loss of generality we may take it to be exactly f. We can now calculate Δq in terms of the derivatives of p, and we have in fact $\Delta q = \Delta p$. Thus the quadratic form becomes $\dfrac{1}{\Delta p}(dp^2 + dq^2)$, and hence x is $1/\Delta p$.

Now if the quadratic form can be written $x(dp^2 + dq^2)$, the surface is said to be divided *isometrically*, or *isothermally*, by the systems of curves p and q, and we have the important result that the solutions of $\Delta_2 p = 0$ give the families of isothermal curves on the surface. We notice that, if one conjugate pair of solutions p, q of $\Delta_2\phi = 0$ has been determined, then taking these as independent variables, we have

$$\Delta_2\phi \equiv \frac{1}{x}\left(\frac{\partial^2\phi}{\partial p^2} + \frac{\partial^2\phi}{\partial q^2}\right),$$

and hence the most general solution of $\Delta_2\phi = 0$ is given by

$$\phi = f_1(p + iq) + f_2(p - iq),$$

where f_1 and f_2 are arbitrary functions of their arguments.

Other examples of this theory (given by Darboux) are

$$ds^2 = \cos^2 x dp^2 + \sin^2 x dq^2,$$

in which case $$\Delta(p, \Delta p) = 2\Delta_2 p(\Delta p - 1),$$

and $$ds^2 = x dp^2 + \frac{1}{x} dq^2,$$

for which $$\Delta(p, \Delta p) = \Delta p\, \Delta_2 p.$$

Examples. (i) Find the relation among the invariants of p if

$$ds^2 = dp^2/\sin^2 x + dq^2/\cos^2 x.$$

(ii) Prove that if the equations $\Delta_2\phi = 0$, $\Delta\left(\frac{\Theta(\phi, \Delta\phi)}{(\Delta\phi)^3}, \phi\right) = 0$, have a common solution, the quadratic can be reduced to Liouville's form,

$$ds^2 = (U + V)(du^2 + dv^2),$$

where U is a function of u only, V a function of v only, and show how to find the parametric curves u, v.

As another illustration, consider the curves ϕ for which

$$\Delta_2\phi/\Delta\phi = f(\phi).$$

If we take these for the parametric curves u, and their orthogonal trajectories for the curves v, we have $ds^2 = A^2du^2 + C^2dv^2$, and

$$\Delta_2 u = \frac{1}{AC}\frac{\partial}{\partial u}\left(\frac{C}{A}\right), \quad \Delta u = \frac{1}{A^2}.$$

Hence our equation becomes

$$\frac{\partial}{\partial u}\left(\log\frac{C}{A}\right) = f(u),$$

and therefore $C = Ae^{U+V}$ where U is a function of u only, V a function of v only. Hence

$$ds^2 = A^2e^{2U}\left[e^{-2U} du^2 + e^{2V} dv^2\right] = \lambda(du'^2 + dv'^2),$$

where u' is a function of u, v' a function of v. The curves u' are the same as the curves ϕ, and it follows that if $\Delta_2\phi/\Delta\phi = f(\phi)$, the curves ϕ form an isothermal system.

58. Applicability of two surfaces.

We now consider the general problem of the applicability of two surfaces. Let the quadratics for the two surfaces be

$$Edu^2 + 2Fdudv + Gdv^2, \quad E_1du_1^2 + 2F_1du_1dv_1 + G_1dv_1^2.$$

We first calculate the invariant K for the two forms, and we must have $K(u, v) = K_1(u_1, v_1)$. This gives one equation for u_1, v_1 in terms of u, v unless K is a constant. As we can have no relation between u, v alone, K_1 must in this case be also the same constant.

Let $K = K_1 = a$, a constant. Choose on the first surface any point P, and take as parametric curves v the geodesics through this point, the curves u being their orthogonal trajectories. We have then for this surface, $ds^2 = du^2 + C^2dv^2$, and C vanishes with u, whatever be the value of v. Similarly, the second quadratic may be reduced to $du_1^2 + C_1^2dv_1^2$. Now $K = -\frac{1}{C}\frac{\partial^2 C}{\partial u^2}$, and therefore $\frac{\partial^2 C}{\partial u^2} + aC = 0$. Hence $C = V\sin\sqrt{a}u$. Similarly $C_1 = V_1\sin\sqrt{a}u_1$. If, then, we take $u = u_1$, $Vdv = V_1dv_1$, the quadratics are transformed into each other, and we see that in this case the surfaces can be applied so that a given point on the first corresponds to any given point on the second, and so that a given geodesic through the point on the first corresponds to any given geodesic through the corresponding point on the second.

Now suppose that K and K_1 are not constants. We must have $\Delta K = \Delta' K_1$, where the index denotes that the invariant is formed with reference to the second of the two forms. There are thus obtained two equations for u_1, v_1 as functions of u, v. If these equations are inconsistent, the surfaces are not applicable. There are two other possibilities; either they are independent of one another, or one is a consequence of the other. In the second case it is easy to see that ΔK must be a function of K, and $\Delta' K_1$ must be the same function of K_1. Suppose that $\Delta K = f(K)$, and form the invariant $\Delta_2 K$. Again there are the three possibilities before mentioned. If

$$\Delta_2 K = \phi(K), \quad \Delta_2'(K_1) = \phi(K_1),$$

take the parametric curves u so that $du = dK/f(K)$. Then $\Delta(u) = 1$,

$\Delta_2(u) = F(u)$, say, and $ds^2 = du^2 + C^2 dv^2$, where the v's are the orthogonal trajectories of the u's. Also $\Delta_2 u = \frac{1}{C}\frac{\partial C}{\partial u}$; hence

$$C = Ve^{\int F(u)\,du},$$

where V is a function of v. Now take $Vdv = dv'$, and it follows that

$$ds^2 = du^2 + e^{\int F(u)\,du}\,dv'^2.$$

Similarly, for the second surface

$$ds^2 = du_1^2 + e^{\int F(u_1)\,du_1}\,dv_1'^2,$$

and the surfaces are deformable into each other by means of $u = u_1$, $v = v_1 + \text{const}$. It thus follows that if

$$K = K_1, \quad \Delta K = \Delta' K_1, \quad \Delta_2 K = \Delta_2' K_1,$$

lead to only one equation between u, v, u_1, v_1, the surfaces are applicable to each other in a single infinity of ways.

Now let the three equations just mentioned lead to two equations between u, v, u_1, v_1. Suppose first that K and ΔK are independent of each other, and take them for the variables in the first form. Similarly, let K_1 and $\Delta' K_1$ be taken for the variables in the second form. Then, if the surfaces are applicable to each other, these two forms must now be the same. It follows at once from the general expression for the form given on p. 56 that we must have

$$\Delta\Delta K = \Delta'\Delta' K_1, \quad \Delta(K, \Delta K) = \Delta'(K_1, \Delta' K_1).$$

If K and ΔK are dependent on one another, we take K and $\Delta_2 K$ as the variables of the form, and the necessary and sufficient conditions become in this case

$$\Delta\Delta_2 K = \Delta'\Delta_2' K_1, \quad \Delta(K, \Delta_2 K) = \Delta'(K_1, \Delta_2' K_1)$$

in addition to $\quad K = K_1, \quad \Delta K = \Delta' K_1, \quad \Delta_2 K = \Delta_2' K_1.$

59. The quadratic form in three variables.

For this case we only note the significance of a few of the more important invariants of lowest order. The geometrical interpretation of those of orders one, two and three for a form of rank zero, the case of ordinary Euclidean space, has been considered in detail by Forsyth in his memoir, already quoted, on the differential invariants of space.

We note that there are six Riemann symbols. The explicit expressions of these have been given by Cayley, and they have been discussed in detail by Lamé for the case in which the parametric surfaces cut orthogonally, in his book on Curvilinear Coordinates. If they all vanish, the space is Euclidean.

The invariant $\Delta\phi \equiv \Sigma a^{(rs)}\phi_r\phi_s$ is $(d\phi/dn)^2$, where dn denotes an element of arc perpendicular to the surface $\phi = \text{const.}$ The invariant

$$\Delta(\phi, \psi) \equiv \Sigma a^{(rs)}\phi_r\psi_s$$

is

$$\left(\frac{d\phi}{dn}\right)\left(\frac{d\psi}{dn_1}\right)\cos\alpha,$$

where α is the angle between the surfaces ϕ, ψ, and dn_1 is normal to ψ.

$$\Theta(\phi, \psi, \chi) \equiv \Sigma\epsilon^{(rst)}\phi_r\psi_s\chi_t$$

is equal to

$$\left(\frac{d\phi}{dn}\right)\left(\frac{d\psi}{dn_1}\right)\left(\frac{d\chi}{dn_2}\right)\sin\Omega,$$

where $\sin\Omega$ is the sine of the solid angle at which the three surfaces ϕ, ψ, χ cut. We thus have the geometric interpretation of all the first order invariants. Suppose that the three functions are taken as u_1, u_2, u_3, the parametric surfaces. Then

$$\Delta(u_r, u_s) = a^{(rs)}.$$

Also if the discriminant of the fundamental form is a,

$$a = \begin{vmatrix} a_{11}, & a_{12}, & a_{13} \\ a_{21}, & a_{22}, & a_{23} \\ a_{31}, & a_{32}, & a_{33} \end{vmatrix},$$

and $a^{(rs)} = \frac{1}{a}A_{rs}$ where A_{rs} is the cofactor of a_{rs} in a. We deduce that the discriminant of the reciprocal form is $1/a$, and $a_{rs} = aA^{(rs)}$, where $A^{(rs)}$ has a similar meaning to A_{rs}. It thus follows that, for example,

$$a_{11} = \{\Delta u_2 . \Delta u_3 - \Delta^2(u_2, u_3)\}\, a,$$

$$a_{12} = \{\Delta(u_2, u_3) . \Delta(u_1, u_3) - \Delta(u_1, u_2) . \Delta u_3\}\, a,$$

where

$$\frac{1}{a} = \begin{vmatrix} \Delta u_1\ , & \Delta(u_1, u_2), & \Delta(u_1, u_3) \\ \Delta(u_1, u_2), & \Delta u_2\ , & \Delta(u_2, u_3) \\ \Delta(u_1, u_3), & \Delta(u_2, u_3), & \Delta u_3 \end{vmatrix},$$

and we thus have the general expression for the quadratic form for any parametric surfaces whatever. We notice that, if we substitute in the determinant for $1/a$ the geometric interpretations of the differential parameters involved, it becomes

$$\left(\frac{du_1}{dn_1}\right)^2\left(\frac{du_2}{dn_2}\right)\left(\frac{du_3}{dn_3}\right)^2\sin^2\Omega,$$

and it is therefore equal to $\Theta^2(u_1, u_2, u_3)$.

Now suppose that $\Delta\phi = f(\phi)$, then, if we take

$$d\phi/\sqrt{f(\phi)} = du,$$

the surfaces $\phi = \text{const.}$ are the same as the surfaces $u = \text{const.}$, and $\Delta u = 1$. If we choose v and w so that

$$\Delta(u, v) = 0, \quad \Delta(u, w) = 0,$$

we have $$ds^2 = du^2 + a_{22}dv^2 + 2a_{23}dvdw + a_{33}dw^2.$$

The square of the element of length for any surface $w = f(v)$ is thus seen to be $du^2 + C^2dv^2$, and hence the curves $dv = 0$, $dw = 0$ are geodesics on this surface. It follows that they are geodesics in the general space. Thus, if $\Delta\phi = f(\phi)$, the surfaces ϕ are the orthogonal trajectories of a normal congruence of geodesic lines.

60. Condition that a family of surfaces form part of a triply orthogonal system.

We next consider the questions: (1) Can the manifold for which ds^2 is the fundamental quadratic contain a triply orthogonal system of surfaces? (2) How is the most general such system to be determined?

It appears that, if the surfaces ϕ const. can form part of a triply orthogonal system, ϕ must satisfy a differential equation of the third order*, and the other two families of surfaces are then determinate. As the method of determining this equation of the third order is an instructive example of the use and advantages of differential invariants, and in particular of covariant differentiation, we give it in detail.

Let u, v, w be three families that form a triply orthogonal system, then

$$P \equiv \Delta(v, w) = 0, \quad Q \equiv \Delta(w, u) = 0, \quad R \equiv \Delta(u, v) = 0.$$

P, Q, R are obviously invariants, and

$$P = \Sigma a^{(rs)} v_r w_s.$$

If we differentiate this covariantively, we have from the rules for covariant differentiation, and since $a^{(rst)} = 0$,

$$P_t = \Sigma a^{(rs)}(v_{rt}w_s + v_r w_{st}), \qquad (rs)$$

* The method here used for the determination of this equation has not previously been given. For the equation itself, for the general quadratic form in three variables, see a note published by the author (*Bull. Amer. Math. Soc.* Vol. XII. p. 379). The equation for the particular case of the form $dx^2 + dy^2 + dz^2$ was first given in explicit form by Cayley, and is obtained by Darboux (*Orthogonal Surfaces*, p. 14 *sqq.*). The reader will find it instructive to compare the method given here for the general case with that of Darboux for the particular case.

where the bracket (rs) denotes that the summation extends to these letters. Also

$$u^{(t)} = \Sigma a^{(pt)} u_p, \qquad (p)$$

and therefore

$$\Sigma P_t u^t = \Sigma a^{(rs)} a^{(pt)} (v_{rt} w_s u_p + v_r w_{st} u_p). \qquad (rspt)$$

Now write $\qquad (v, wu) \equiv \Sigma a^{(rs)} a^{(pt)} v_{rt} w_s u_p \qquad (rspt)$

then (v, wu) is an invariant, and we have

$$(v, wu) + (w, uv) = \Sigma P_t u^{(t)}.$$

Also $(v, wu) = (v, uw)$, since $v_{rt} = v_{tr}$. (It happens that for second derivatives the order of differentiation does not matter, though this is not the case for higher derivatives.) We thus have two similar equations, and hence

$$2(u, vw) = \Sigma (Q_t v^{(t)} + R_t w^{(t)} - P_t u^{(t)}). \qquad (t)$$

Now since P, Q, R are invariants, P_t, Q_t, R_t are ordinary derivatives, and hence if P, Q, R are zero so also are these derivatives, and therefore

$$(u, vw) = 0.$$

This equation is thus a consequence of our first three. It involves only first derivatives of v and w, and in fact it is linear and homogeneous in quantities of the type $v_p w_p$, $v_p w_q + v_q w_p$. In this respect it is similar to $P = 0$; it however involves first and second derivatives of u. We may write it

$$(u, vw) \equiv I \equiv \Sigma A^{(pq)} v_p w_q = 0, \qquad (pq)$$

then the A's form a contravariant system of the second order, and they are functions of u and its first and second derivatives; also

$$A^{(pq)} = A^{(qp)}.$$

If we form the covariant derivative of I we have

$$I_h = \Sigma A^{(pq)} (v_{ph} w_q + v_p w_{qh}) + \Sigma A^{(pqk)} a_{hk} v_p w_q, \qquad (pqk)$$

and hence, as in the previous case,

$$\Sigma I_h u^{(h)} = \Sigma A^{(pq)} a^{(ht)} (v_{ph} w_q + v_p w_{qh}) u_t + \Sigma A^{(pqk)} a^{(ht)} a_{hk} v_p w_q u_t. \quad (pqhkt)$$

Also $\qquad R_p = \Sigma a^{(ht)} v_{hp} u_t + \Sigma a^{(ht)} v_h u_{pt}, \qquad (ht)$

$$Q_q = \Sigma a^{(ht)} w_{hq} u_t + \Sigma a^{(ht)} w_h u_{qt}, \qquad (ht)$$

and therefore we can at once eliminate the second derivatives of v and w from the above equation. It then becomes

$$\Sigma I_h u^{(h)} = \Sigma A^{(pq)} w_q (R_p - \Sigma a^{(ht)} u_{pt} v_h) + \Sigma A^{(pq)} v_p (Q_q - \Sigma a^{ht} u_{qt} w_h)$$
$$+ \Sigma A^{(pqk)} a^{(ht)} a_{hk} u_t v_p w_q, \quad (pqkht)$$

or, if we bring over the terms in R and Q to the left, and change the letters of summation so as to have $v_p w_q$ in the typical term in each sum on the right, we have

$$\Sigma I_h u^{(h)} - \Sigma A^{(pq)} R_p w_q - \Sigma A^{(pq)} v_p Q_q = \Sigma A^{(pqk)} a^{(ht)} a_{hk} u_t v_p w_q$$
$$- \Sigma A^{(hq)} a^{(pt)} u_{ht} v_p w_q - \Sigma A^{(ph)} a^{(qt)} u_{ht} v_p w_q \quad (pqkht)$$
$$= \Sigma B^{(pq)} v_p w_q, \quad (pq)$$

where the coefficients $B^{(pq)}$ form a contravariant system of the second order, and

$$B^{(pq)} = B^{(qp)} = \Sigma A^{(pqk)} a^{(ht)} a_{hk} u_t - \Sigma u_{ht} (a^{(pt)} A^{(hq)} + a^{(qt)} A^{(hp)}). \quad (kht)$$

Now, since $I = 0$, $R = 0$, $Q = 0$, it follows as before that I_h, R_p, Q_q are all zero, and hence the left side of the equation just given vanishes. It therefore follows that

$$\Sigma B^{(pq)} v_p w_q = 0. \quad (pq)$$

This equation is thus another consequence of the first three. It involves only first derivatives, in the combinations $v_p w_p$, $v_p w_q + v_q w_p$, of the two functions v, w. Its coefficients are, however, functions involving the third derivatives of u.

We now have three equations

$$\Sigma a^{(rs)} v_r w_s, \quad \Sigma A^{(rs)} v_r w_s, \quad \Sigma B^{(rs)} v_r w_s,$$

linear and homogeneous in the six quantities

$$v_1 w_1, \quad v_2 w_2, \quad v_3 w_3, \quad v_2 w_3 + v_3 w_2, \quad v_3 w_1 + v_1 w_3, \quad v_1 w_2 + v_2 w_1,$$

and, in addition,

$$\Delta (u, v) \equiv \Sigma a^{(rs)} u_r v_s \equiv \Sigma u^{(s)} v_s = 0,$$
$$\Delta (u, w) \equiv \Sigma u^{(s)} w_s = 0.$$

Now from

$$u^{(1)} v_1 + u^{(2)} v_2 + u^{(3)} v_3 = 0, \quad u^{(1)} w_1 + u^{(2)} w_2 + u^{(3)} w_3 = 0,$$

we deduce by multiplying the former by w_1, the latter by v_1, and adding,

$$2u^{(1)} v_1 w_1 + u^{(2)} (v_1 w_2 + v_2 w_1) + u^{(3)} (v_3 w_1 + v_1 w_3) = 0,$$

which is another equation linear in the six quantities above mentioned. We can similarly obtain two other such equations, and now have altogether six equations, from which we can eliminate the derivatives of v and w; the final result is

$$\begin{vmatrix} B^{(11)}, & B^{(22)}, & B^{(33)}, & B^{(23)}, & B^{(31)}, & B^{(12)} \\ A^{(11)}, & A^{(22)}, & A^{(33)}, & A^{(23)}, & A^{(31)}, & A^{(12)} \\ a^{(11)}, & a^{(22)}, & a^{(33)}, & a^{(23)}, & a^{(31)}, & a^{(12)} \\ 2u^{(1)}, & 0\ , & 0\ , & 0\ , & u^{(3)}, & u^{(2)} \\ 0\ , & 2u^{(2)}, & 0\ , & u^{(3)}, & 0\ , & u^{(1)} \\ 0\ , & 0\ , & 2u^{(3)}, & u^{(2)}, & u^{(1)}, & 0 \end{vmatrix} = 0,$$

which is the differential equation of the third order for u. This equation may also be written

$$\Sigma \epsilon_{abc} \epsilon_{def} \epsilon_{ghk} A^{(ad)} B^{(bg)} a^{(eh)} u^{(c)} u^{(f)} u^{(k)} = 0, \qquad (abcdefghk)$$

and this shows that the above determinant, which is $2a^{-\frac{3}{2}}$ times the left side of this equation, is an invariant. It is in fact the algebraic invariant of four ternary forms, three quadratic and one linear, which is linear in the coefficients of each of the quadratic forms, and cubic in the coefficients of the linear form.

This result might have been obtained by contravariant differentiation, starting with the expression $\sum_{rs} a_{rs} u^{(r)} v^{(s)}$ for $\Delta(u, v)$; the work would have been of the same character and the final result of the same form, the only difference being that each contravariant derivative involved would be replaced by its reciprocal covariant derivative with reference to the fundamental form.

61. The quadratic form in n variables.

A general method for dealing with a manifold for which ds^2 is given is due to Ricci and Levi Civita. The variables being $x_1, \ldots, x_n$, they write

$$dx_r/ds = \lambda^{(r)}. \qquad (r = 1, 2, \ldots, n)$$

Then $\lambda^{(1)}, \ldots, \lambda^{(n)}$ is a contravariant system of the first order, and in fact $\sum_{r,s} a_{rs} \lambda^{(r)} \lambda^{(s)} = 1$. The reciprocal system $\lambda_1, \ldots, \lambda_n$ is given by $\lambda_s = \sum_r a_{rs} \lambda^{(r)}$, and therefore $\sum_s \lambda^{(s)} \lambda_s = 1$, and also $\sum_{r,s} a^{(rs)} \lambda_r \lambda_s = 1$. If we suppose the quantities $\lambda^{(r)}$ arbitrarily given functions of the independent variables x, the values of the ratios dx_r/ds are given for any set of values of these variables. There is thus associated a direction with each point in the manifold, and, if we start from a given point and proceed always in the direction associated with the point reached, we finally obtain a curve in the manifold. Hence the equations given above define a congruence of curves in the manifold, such that there is one and only one curve through each point of the manifold. Let μ define another such congruence, then the cosine of the angle between two intersecting curves, one belonging to the congruence λ, and the other to the congruence μ, may be proved to be

$$\sum_r \lambda^{(r)} \mu_r = \sum_r \lambda_r \mu^{(r)} = \sum_{r,s} a_{rs} \lambda^{(r)} \mu^{(s)} = \sum_{r,s} a^{(rs)} \lambda_r \mu_s.$$

If in particular the curves of the one congruence are everywhere orthogonal to the curves of the other, $\sum_r \lambda^{(r)} \mu_r = 0$.

62. Orthogonal ennuple.

Now take n such congruences, denoted by $\lambda_{h/r}$ or $\lambda_h^{(r)}$, ($h=1, 2, \ldots, n$), where the letter h denotes the congruence. (λ_h is one symbol and is not a covariant derivative.) Let it be assumed that all the curves of these congruences cut each other orthogonally, then

$$\sum_r \lambda_h^{(r)} \lambda_{k/r} = \eta_{hk}, \qquad (h, k = 1, 2, \ldots, n)$$

where $\eta_{hk} = 0$ if $h \neq k$, and $\eta_{hh} = 1$. Such a set of congruences is called an *orthogonal ennuple*; the notation [1], [2], ..., [n] is used for the congruences of the ennuple, 1, 2, ..., n are the curves of the congruence that pass through a given point, and $s_1, s_2, \ldots, s_n$ are arcs of these curves.

63. Any covariant or contravariant system can be expressed in terms of invariants and the coefficients of an orthogonal ennuple.

For let $X_{r_1 \ldots r_m}$ be a member of such a system, then

$$c_{h_1 h_2 \ldots h_m} \equiv \sum_{r_1 \ldots r_m} \lambda_{h_1}^{(r_1)} \ldots \lambda_{h_m}^{(r_m)} X_{r_1 \ldots r_m},$$

is an invariant, and on solving all the equations of this type for the X's we have

$$X_{r_1 \ldots r_m} = \sum_{h_1 \ldots h_m} c_{h_1 \ldots h_m} \lambda_{h_1/r_1} \ldots \lambda_{h_m/r_m},$$

which proves the theorem. It follows that if the members of the system X are all zero, the invariants c are all zero. Thus any absolute system of equations may be expressed by the vanishing of a set of invariants.

A particular case of the theorem just proved is that the covariant derivatives of the λ's may be expressed in terms of the λ's themselves. Thus let

$$\gamma_{hkl} = \sum_{r, s} \lambda_k^{(r)} \lambda_l^{(s)} \lambda_{h/rs}, \quad \text{then} \quad \lambda_{h/rs} = \sum_{k, l} \gamma_{hkl} \lambda_{k/r} \lambda_{l/s}.$$

Now there exist relations among the λ's, and therefore there are relations among their covariant derivatives, which are obtained by differentiating covariantively the equations of type

$$\sum_r \lambda_h^{(r)} \lambda_{k/r} = \eta_{hk}.$$

We thus obtain

$$\sum_r \lambda_h^{(r)} \lambda_{k/rs} + \sum_r \lambda_h^{(rt)} a_{ts} \lambda_{k/r} = 0,$$

or

$$\sum_r \lambda_h^{(r)} \lambda_{k/rs} + \sum_r \lambda_{h/rs} \lambda_k^{(r)} = 0.$$

On substituting in this the values of the quantities $\lambda_{h/rs}$, $\lambda_{k/rs}$, we obtain

$$\sum_{r,\,p,\,q} \gamma_{hpq}\lambda_{p/r}\lambda_{q/s}\lambda_k^{(r)} + \sum_{r,\,p,\,q} \gamma_{kpq}\lambda_{p/r}\lambda_{q/s}\lambda_h^{(r)} = 0,$$

or

$$\sum_q \gamma_{hkq}\lambda_{q/s} + \sum_q \gamma_{khq}\lambda_{q/s} = 0,$$

for all values of s. It therefore follows that $\gamma_{hkq} + \gamma_{khq} = 0$, and in particular $\gamma_{hhq} = 0$. Thus the number of independent invariants γ is

$$\tfrac{1}{2}n^2(n-1).$$

64. Coefficients of rotation of an ennuple.

It is clear, from what has been said, that the geometrical properties of the ennuple are all contained in the invariants γ. These invariants are called the coefficients of rotation of the ennuple.

65. Let f be any function of the variables, then $\partial f/\partial s_h$ is an invariant, and it is equal to $\sum_r \lambda_h^{(r)} f_r$. ($\partial f/\partial s_h$ is merely the ratio of two infinitesimal increments, and not a derivative, for f cannot be regarded as a function of n quantities $s_1, s_2, \ldots, s_n$.) Since this is an invariant, we obtain by covariant differentiation

$$\frac{\partial}{\partial x_p}\frac{\partial f}{\partial s_h} = \sum_r \lambda_h^{(r)} f_{rp} + \sum_r f^{(r)}\lambda_{h/rp},$$

and consequently

$$\frac{\partial}{\partial s_k}\frac{\partial f}{\partial s_h} \equiv \sum_p \lambda_k^{(p)}\frac{\partial}{\partial x_p}\frac{\partial f}{\partial s_h} = \sum_{r,\,p} \lambda_h^{(r)}\lambda_k^{(p)} f_{rp} + \sum_{r,\,p} f^{(r)}\lambda_k^{(p)}\lambda_{h/rp},$$

or on substitution for the quantities $\lambda_{h/rp}$,

$$\frac{\partial}{\partial s_k}\frac{\partial f}{\partial s_h} = \sum_{r,\,p} \lambda_h^{(r)}\lambda_k^{(p)} f_{rp} + \sum_i \gamma_{hik}\frac{\partial f}{\partial s_i},$$

and therefore

$$\frac{\partial}{\partial s_k}\frac{\partial f}{\partial s_h} - \frac{\partial}{\partial s_h}\frac{\partial f}{\partial s_k} = \sum_i (\gamma_{ikh} - \gamma_{ihk})\frac{\partial f}{\partial s_i}.$$

66. Normal congruence.

By means of this identity we can determine the condition that a congruence may be *normal.* A congruence is said to be normal if its curves can be considered as the orthogonal trajectories of a family of surfaces $f(x_1, \ldots, x_n) = \text{const.}$ Suppose that $[n]$ is a normal congruence, having the surfaces f as orthogonal trajectories. Then any curve of the congruences $[1], [2], \ldots, [n-1]$ must lie in a surface f, and therefore

$$\frac{\partial f}{\partial s_h} = 0. \qquad (h = 1, 2, \ldots, \overline{n-1})$$

Also $\dfrac{\partial f}{\partial s_n}$ cannot be zero, for the curves $[n]$ are perpendicular to f. Hence, if we substitute in the above relation, we deduce that

$$\gamma_{nkh} = \gamma_{nhk},$$

where h, k take all values from 1 to $n-1$. These conditions are also sufficient, for if they are satisfied

$$\frac{\partial}{\partial s_k}\frac{\partial F}{\partial s_h} - \frac{\partial}{\partial s_h}\frac{\partial F}{\partial s_k}$$

is expressed linearly in terms of the quantities $\dfrac{\partial F}{\partial s_i}$, where i, h, k take all values from 1 to $n-1$. This is the condition that the system of equations $\dfrac{\partial F}{\partial s_i} = 0$ may have a common solution. If all the congruences are normal we have relations $\gamma_{rkh} = \gamma_{rhk}$, for all values of r from 1 to n, where h, k take all values from 1 to n excluding r, and also

$$\gamma_{hkl} + \gamma_{khl} = 0$$

for all values of h, k, l. It therefore follows that all the γ's with three different suffixes are zero; reciprocally, if all the γ's with three different suffixes are zero, the ennuple is normal.

Suppose that $[n]$ is normal; then

$$\sum_r \lambda_h{}^{(r)} f_r = 0. \qquad (h = 1, 2, \ldots, n-1)$$

Also

$$\sum_r \lambda_h{}^{(r)} \lambda_{n/r} = 0, \qquad (h = 1, 2, \ldots, n-1)$$

and therefore f_r is proportional to $\lambda_{n/r}$. Let $f_r = \mu\lambda_{n/r}$, then since $f_{rs} = f_{sr}$ and

$$f_{rs} = \mu_s \lambda_{n/r} + \mu \sum_{ij} \gamma_{nij} \lambda_{i/r} \lambda_{j/s},$$

we deduce that

$$\mu_s \lambda_{n/r} + \mu \sum_{ij} \gamma_{nij} \lambda_{i/r} \lambda_{j/s} = \mu_r \lambda_{n/s} + \mu \sum_{ij} \gamma_{nij} \lambda_{i/s} \lambda_{j/r}.$$

If we multiply this equation by $\lambda_n{}^{(s)}$ and sum over all values of s from 1 to n, we get

$$\sum_s \mu_s \lambda_n{}^{(s)} \lambda_{n/r} + \mu \sum_{i=1}^{n-1} \gamma_{nin} \lambda_{i/r} = \mu_r,$$

or, finally, if

$$\sum_s \mu_s \lambda_n{}^{(s)} \equiv \mu\nu,$$

$$\frac{\mu_r}{\mu} = \nu\lambda_{n/r} + \sum_{i=1}^{n-1} \gamma_{nin} \lambda_{i/r}.$$

67. Families of isothermal surfaces.

A family of surfaces f is said to be isothermal if

$$\Sigma a^{(rs)} f_{rs} = 0.$$

Such families are the generalisation of equipotential surfaces in ordinary space. An isothermal family may be considered given if its normal congruence is given; let then $[n]$ be its normal congruence. If we substitute the value of f_{rs} in terms of the λ's in the equation

$$\Sigma a^{(rs)} f_{rs} = 0,$$

we have
$$\sum_{r,s} \mu_s a^{(rs)} \lambda_{n/r} + \mu \sum_{ijrs} \gamma_{nij}\, a^{(rs)}\, \lambda_{i/r}\, \lambda_{j/s} = 0,$$

or
$$\sum_s \mu_s \lambda_n^{(s)} + \mu \sum_i \gamma_{nii} = 0,$$

that is to say
$$\nu = -\sum_{i=1}^{n-1} \gamma_{nii}.$$

If, then, $[n]$ is normal to an isothermal family, this value of ν must make

$$\sum_{r=1}^{n} \{\nu \lambda_{n/r} + \sum_{i=1}^{n-1} \gamma_{nin} \lambda_{i/r}\}\, dx_r,$$

a perfect differential. The conditions for this are

$$\frac{\partial \nu}{\partial s_h} + \frac{\partial \gamma_{hnn}}{\partial s_n} + \nu \gamma_{hnn} + \sum_{i=1}^{n-1} \gamma_{inn} (\gamma_{ihn} - \gamma_{inh}) = 0,$$

$$\frac{\partial \gamma_{hnn}}{\partial s_k} + \sum_{i=1}^{n-1} \gamma_{inn} \gamma_{ihk} = \frac{\partial \gamma_{knn}}{\partial s_h} + \sum_{i=1}^{n-1} \gamma_{inn} \gamma_{ikh},$$

$$(h, k = 1, 2, \ldots, n-1).$$

These reduce to a much simpler form if the ennuple is normal.

68. Geodesic congruences.

Let the coordinates of a point on any curve in the manifold be expressed as functions of a parameter t, and let accents denote derivatives with regard to t, then for this curve

$$s'^2 = \sum_{p,q} a_{pq} x_p' x_q',$$

and therefore the length of the arc joining two points whose parameters are t_0 and t_1 is

$$s \equiv \int_{t_0}^{t_1} s' dt \equiv \int_{t_0}^{t_1} \sqrt{\Sigma a_{pq} x_p' x_q'}\, dt.$$

Let δ denote a variation due to taking a neighbouring curve, then

$$2s' \delta s' = 2 \sum_{p,q} a_{pq} x_p' \delta x_q' + \sum_{p,q,k} \frac{\partial a_{pq}}{\partial x_k} x_p' x_q' \delta x_k.$$

If the curve we are considering belongs to the congruence λ, then

$$x_p' = s'\lambda^{(p)}, \quad \sum_p a_{pq}\,\lambda^{(p)} = \lambda_q\,;$$

also

$$[pq, k] + [pk, q] = \frac{\partial a_{qk}}{\partial x_p}.$$

When these values are substituted in the above expression for $\delta s'$ and the result divided through by $2s'$, there is obtained the equation

$$\delta s' = \sum_p \lambda_p \partial x_p' + s' \sum_k \delta x_k \sum_{p,q} [kp, q]\,\lambda^{(p)}\lambda^{(q)}.$$

Now from the definition for covariant derivatives we have

$$\lambda_{hk} = \frac{\partial}{\partial x_k}\lambda_h - \sum_q \{hk, q\}\,\lambda_q,$$

also

$$\{hk, q\} = \sum_r a^{(qr)}\,[hk, r].$$

Hence, since

$$\sum_q a^{(qr)}\lambda_q = \lambda^{(r)},$$

it follows that

$$\lambda_{hk} = \frac{\partial}{\partial x_k}\lambda_h - \sum_r [hk, r]\,\lambda^{(r)},$$

and therefore, changing the letters of summation, we have

$$\sum_p \lambda^{(p)}\lambda_{kp} = \sum_p \lambda^{(p)}\frac{\partial}{\partial x_p}\lambda_k - \sum_{p,q} [kp, q]\,\lambda^{(p)}\lambda^{(q)}.$$

If we substitute from this the quantity

$$\sum_{p,q} [kp, q]\,\lambda^{(p)}\lambda^{(q)}$$

in the expression for $\delta s'$, we obtain

$$\delta s' = \sum_p \lambda_p \delta x_p' + s' \sum_{p,k} \lambda^{(p)}\,\delta x_k \frac{\partial \lambda_k}{\partial x_p} - s' \sum_k \delta x_k \sum_p \lambda^{(p)}\lambda_{kp},$$

$$= \frac{d}{dt}\left[\sum_p \lambda_p\,\delta x_p\right] - s' \sum_k \delta x_k \sum_p \lambda^{(p)}\,\lambda_{kp}.$$

Hence, on integration

$$\delta s = \left[\sum_p \lambda_p \delta x_p\right]_{t_0}^{t_1} - \int_{t_0}^{t_1} s' \sum_k \delta x_k \sum_p \lambda^{(p)}\,\lambda_{kp}\,dt.$$

If this variation is zero for all possible neighbouring curves, we must have, in addition to conditions at the limits,

$$\sum_k \delta x_k \sum_p \lambda^{(p)}\lambda_{kp} = 0,$$

at every point on the curve, for all possible values of the δx's. Hence,

$$\sum_p \lambda^{(p)}\,\lambda_{kp} = 0$$

for all values of k. A curve for which this first variation δs is zero is called a geodesic. It is clear that if s is a minimum, δs must be zero, and it may be shown that, subject to certain conditions of regularity, and provided the length of the arc s is sufficiently small, if δs is zero s is a minimum.

The congruence λ is therefore composed of geodesic lines if

$$\sum_p \lambda^{(p)} \lambda_{kp} = 0. \qquad (k = 1, 2, \ldots, n)$$

Suppose that the congruence $[n]$ is geodesic. Then since

$$\lambda_{n/kp} = \sum_{ij} \gamma_{nij} \lambda_{i/k} \lambda_{j/p},$$

we have

$$\sum_{pij} \gamma_{nij} \lambda_n^{(p)} \lambda_{i/k} \lambda_{j/p} = 0,$$

or

$$\sum_i \gamma_{nin} \lambda_{i/k} = 0. \qquad (k = 1, 2, \ldots, n)$$

From this it follows that all the quantities γ_{nin} $(i = 1, 2, \ldots, n)$ are zero. Thus *if* $[n]$ *is geodesic all the invariants* γ_{nin} *are zero, and conversely.* In particular, if the manifold determined by the quadratic form is Euclidean, these are the conditions that all the curves of the congruence may be straight lines.

In the general case, if the congruence is also normal to the surface f,

$$f_r = \mu \lambda_{n/r}, \quad \mu_r/\mu = \nu \lambda_{n/r}.$$

Hence $f_r/\mu_r = f_s/\mu_s$ for all the values of r and s, and therefore μ is a function of f. Hence we may modify the function f so as to have exactly $f_r = \lambda_{n/r}$ for all values of r.

69. Congruences canonical with reference to a given congruence.

In many problems associated with a quadratic form, it happens that one congruence is given. If this is $[n]$, the $n-1$ other congruences that form with it an orthogonal ennuple may be chosen in an infinity of ways. Out of this infinite number of systems of $n-1$ congruences, there is one (or more) for which many of the relations are simplified. We proceed to define a set of $n-1$ congruences which are called *canonical* with reference to the congruence $[n]$. Write

$$2X_{rs} = \lambda_{n/rs} + \lambda_{n/sr},$$

and consider the system of algebraic equations

$$\sum_{r=1}^{n} \lambda_{n/r} \lambda^{(r)} = 0,$$

$$\lambda_{n/q} \mu + \sum_{r=1}^{n} (X_{qr} + \omega a_{qr}) \lambda^{(r)} = 0, \quad (q = 1, 2, \ldots, n)$$

for the unknown quantities $\lambda^{(1)}, \ldots, \lambda^{(n)}, \omega, \mu$. These equations are linear in μ and the λ's, when these quantities are eliminated we have an equation $\Delta(\omega) = 0$ of degree $n-1$ for ω. Suppose that the roots of Δ are all simple. Then to any root, say ω_h, there corresponds a set of values for the λ's. We thus have $n-1$ congruences which are obviously orthogonal to [n]; they are orthogonal to each other. For, multiplying the second of the above equations by $\lambda_k^{(q)}$ and summing for all values of q, we have

$$\sum_q \lambda_k^{(q)} \lambda_{n/q} \mu + \sum_{q, r} (X_{qr} + \omega_h a_{qr}) \lambda_h^{(r)} \lambda_k^{(q)} = 0,$$

or

$$\sum_{q, r} (X_{qr} + \omega_h a_{qr}) \lambda_h^{(r)} \lambda_k^{(q)} = 0.$$

Similarly

$$\sum_{q, r} (X_{qr} + \omega_k a_{qr}) \lambda_h^{(r)} \lambda_k^{(q)} = 0,$$

and therefore

$$(\omega_h - \omega_k) \sum a_{qr} \lambda_h^{(r)} \lambda_k^{(q)} = 0,$$

and since $\omega_h \neq \omega_k$, [h] is orthogonal to [k]. The set of congruences thus determined is called *canonical* with reference to [n]. If the roots of $\Delta(\omega)$ are not all simple, suppose that one root is of multiplicity p. Then we choose p congruences associated with this root and orthogonal among themselves, and similarly, by taking account of all the roots, we can obtain a canonical set in this special case. The set is not, however, definite; in the p congruences associated with the root above mentioned, for example, there is the same amount of arbitrariness as there is in a linear orthogonal substitution on p variables.

It is easy to prove that, for an orthogonal ennuple in which [1], [2], ..., [$n-1$] form the canonical system with reference to [n],

$$\gamma_{nhk} + \gamma_{nkh} = 0, \qquad (h, k = 1, 2, \ldots, n-1; \ h \neq k)$$

and also

$$\omega_k = -\gamma_{nkk}.$$

If [n] is normal, it follows that, with the same restrictions on h, k, $\gamma_{nhk} = 0$. In this case the $n-1$ congruences are the $n-1$ families of lines of curvature of the surfaces orthogonal to the curves [n].

70. Geometrical interpretation of the invariants γ.

The invariants γ of an orthogonal ennuple have an important geometrical interpretation. This will be illustrated by a simple case. We shall suppose the quadratic form to be that for ordinary Euclidean space, that is to say, it involves three variables, and the six Riemann symbols are all zero. In this case there are three mutually orthogonal congruences [1], [2], [3]. Now consider a frame of axes given by the lines drawn through a fixed point parallel to the tangents at any point to the curves 1, 2, 3 through that point. If we suppose the quadratic to be $dx^2 + dy^2 + dz^2$, the direction cosines of the lines 1, 2, 3 are given by the scheme

	1	2	3
x	$\lambda_{1/1}$	$\lambda_{2/1}$	$\lambda_{3/1}$
y	$\lambda_{1/2}$	$\lambda_{2/2}$	$\lambda_{3/2}$
z	$\lambda_{1/3}$	$\lambda_{2/3}$	$\lambda_{3/3}$

If $x + dx$, $y + dy$, $z + dz$ is a point consecutive to x, y, z, then it is easily seen that

$$dx = \sum_p \lambda_{p/1}\, ds_p, \quad dy = \sum_p \lambda_{p/2}\, ds_p, \quad dz = \sum_p \lambda_{p/3}\, ds_p.$$

Also, in this case, $\lambda_p^{(r)} = \lambda_{p/r}$ and $\lambda_{h/rs} = \dfrac{\partial}{\partial x_s}\lambda_{h/r}$, where $x_1, x_2, x_3 \equiv x, y, z$. Again (Darboux, *Theory of Surfaces*, Vol. I. p. 5) if p, q, r be the infinitesimal rotations whereby the axes are brought to their consecutive position,

$$\begin{aligned} p &= \sum_r \lambda_{3/r}\, d\lambda_{2/r} \\ &= \sum_{r,\, s} \lambda_{3/r}\lambda_{2/rs}\, dx_s \\ &= \sum_{r,\, s,\, p} \lambda_{3/r}\lambda_{2/rs}\lambda_{p/s}\, ds_p \\ &= \sum_{r,\, s,\, p} \lambda_3^{(r)}\lambda_p^{(s)}\lambda_{2/rs}\, ds_p \\ &= \sum_p \gamma_{23p}\, ds_p. \end{aligned}$$

Similarly $$q = \sum_p \gamma_{31p}\, ds_p, \quad r = \sum_p \gamma_{12p}\, ds_p,$$

and hence the γ's are seen to be the coefficients of rotation of the frame of axes.

71. Relations among the derivatives of the γ's.

Of the γ's associated with a general form there are $\frac{1}{2}n^2(n-1)$ algebraically independent. These are not however absolutely independent.

In fact it may be proved by forming the second covariant derivatives of the λ's that if

$$\gamma_{hi,\,kl} \equiv \frac{\partial}{\partial s_l}\gamma_{hik} - \frac{\partial}{\partial s_k}\gamma_{hil} + \sum_j \{\gamma_{hij}(\gamma_{jkl} - \gamma_{jlk}) + \gamma_{jhl}\gamma_{jik} - \gamma_{jhk}\gamma_{jil}\},$$

then
$$\gamma_{hi,\,kl} = \sum_{q,\,r,\,s,\,t} \lambda_h^{(q)}\lambda_i^{(r)}\lambda_k^{(s)}\lambda_l^{(t)}\,(qrst).$$

These are the generalisation of the relations given by Darboux (*Theory of Surfaces*, Vol. I. Chap. v) between the rotations of a system of axes in ordinary Euclidean space. If $n = 2$, they reduce to the single relation

$$\frac{\partial}{\partial s_2}\gamma_{121} = \frac{\partial}{\partial s_1}\gamma_{212} = \gamma_{121}^2 + \gamma_{212}^2 + K.$$

This has an immediate geometrical interpretation, for γ_{121} and γ_{212} are the geodesic curvatures of the curves 1 and 2.

72. Condition that a quadratic form admits an infinitesimal transformation.

As an application of the theory of congruences given above, we consider the following question: What are the invariant relations that must be satisfied in order that a manifold for which the quadratic form is given may be transformable into itself by an infinitesimal transformation? In other words: What are the conditions that a given quadratic form may admit an infinitesimal transformation?

An example of a quadratic form which admits an infinitesimal transformation is given by ds^2 for a surface of revolution in ordinary space, for such a surface is transformed into itself by an infinitesimal rotation about its axis.

In the general case, let the infinitesimal transformation be

$$Xf \equiv \sum_r \xi^{(r)}\frac{\partial f}{\partial x_r},$$

then, if $(dF/dt)\,dt$ denotes the change in any quantity F due to the transformation,

$$\frac{dx_r}{dt} = \xi^{(r)}, \quad \frac{d}{dt}(dx_r) = d\xi^{(r)} = \sum_p \frac{\partial \xi^{(r)}}{\partial x_p}dx_p, \quad \frac{da_{rs}}{dt} = \sum_p \xi^{(p)}\frac{\partial a_{rs}}{\partial x_p},$$

and therefore the change in the quadratic form is

$$\left\{\sum_{p,\,r,\,s}\xi^{(p)}\frac{\partial a_{rs}}{\partial x_p}dx_r dx_s + 2\sum_{r,\,s,\,p} a_{rs}\frac{\partial \xi^{(s)}}{\partial x_p}dx_p dx_r\right\}dt.$$

The necessary and sufficient condition that the quadratic form may admit the infinitesimal transformation is that this must be identically zero. Now
$$\xi_r = \sum_s a_{rs}\xi^{(s)},$$

and therefore $$\frac{\partial \xi_r}{\partial x_p} = \sum_s a_{rs} \frac{\partial \xi^{(s)}}{\partial x_p} + \sum_s \xi^{(s)} \frac{\partial a_{rs}}{\partial x_p}.$$

Also $$\xi_{rp} = \frac{\partial \xi_r}{\partial x_p} - \sum_q \{rp, q\} \xi_q = \frac{\partial \xi_r}{\partial x_p} - \sum_t [rp, t] \xi^{(t)},$$

and it therefore follows that

$$\sum_s a_{rs} \frac{\partial \xi^{(s)}}{\partial x_p} = \xi_{rp} + \sum_t [rp, t] \xi^{(t)} - \sum_s \xi^{(s)} \frac{\partial a_{rs}}{\partial x_p}.$$

Again, $[pk, q] + [qk, p] = \frac{\partial a_{pq}}{\partial x_k}$, and hence the condition becomes

$$\sum_{p, r, s} \xi^{(p)} ([rp, s] + [sp, r]) \, dx_r dx_s + 2 \sum_{r, p} \{\xi_{rp} + \sum_t [rp, t] \xi^{(t)} - \sum_s \xi^{(s)} ([rp, s] + [sp, r])\} \, dx_r dx_p = 0,$$

or finally $$\sum_{r, p} \xi_{rp} \, dx_r dx_p = 0.$$

Hence the necessary and sufficient conditions that the quadratic form may admit the infinitesimal transformation are

$$\xi_{rs} + \xi_{sr} = 0,$$

for all values of r and s. (This result is due to Killing, *Crelle*, Vol. 109.)

The quantities ξ_r determine a congruence of curves in the manifold; these are the paths of points in the displacement. Let us write $\xi_r = \rho \lambda_{n/r}$; then

$$\lambda_{n/rs} = \sum_{ij} \gamma_{nij} \lambda_{i/r} \lambda_{j/s},$$

and it follows that

$$\xi_{rs} = \sum_{ij} \rho \gamma_{nij} \lambda_{i/r} \lambda_{j/s} + \rho_s \lambda_{n/r}.$$

Hence since $\xi_{rs} + \xi_{sr} = 0$,

$$\sum_{ij} \rho (\gamma_{nij} + \gamma_{nji}) \lambda_{i/r} \lambda_{j/s} + \rho_s \lambda_{n/r} + \rho_r \lambda_{n/s} = 0. \quad (r, s = 1, 2, \ldots, n)$$

We multiply this equation by $\lambda_k^{(s)}$ and sum over the values of s, and thus obtain the equivalent set of equations

$$(a) \quad \sum_i \rho (\gamma_{nik} + \gamma_{nki}) \lambda_{i/r} + b_k \lambda_{n/r} = 0, \quad (k = 1, 2, \ldots, n-1)$$

$$(b) \quad \sum_i \rho (\gamma_{nin} + \gamma_{nni}) \lambda_{i/r} + b_n \lambda_{n/r} + \rho_r = 0,$$

where b_k is written for $\sum_s \rho_s \lambda_k^{(s)}$.

Again, multiply (a) by $\lambda_h^{(r)}$ and the sum over the values of r; the results are

$$\gamma_{nhk} + \gamma_{nkh} = 0, \quad (h, k = 1, 2, \ldots, n-1)$$

and $$\rho (\gamma_{nnk} + \gamma_{nkn}) + b_k = 0. \quad (k = 1, 2, \ldots, n-1)$$

Similarly from (b) we obtain the additional equation $b_n = 0$. Also $\gamma_{nnk} = 0$ for all values of k, for any orthogonal ennuple whatever, and therefore the necessary and sufficient conditions that the quadratic form may admit the given infinitesimal transformation are, finally,

$$(A) \quad \gamma_{nhk} + \gamma_{nkh} = 0, \qquad (h, k = 1, 2, \ldots, n-1)$$

$$(B) \quad \rho\gamma_{nhn} + b_h = 0, \qquad (h = 1, 2, \ldots, n-1)$$

$$(C) \quad b_n = 0,$$

where $b_h = \sum_s \rho_s \lambda_h^{(s)}$.

73. These equations are in invariative form, and we proceed to interpret them geometrically. In the first place the set of equations (A) include in them all the conditions satisfied by a canonical system of congruences with reference to $[n]$; thus every system of congruences forming with $[n]$ an orthogonal ennuple is canonical with reference to $[n]$. The remaining equations included in (A) are

$$\gamma_{nhh} = 0. \qquad (h = 1, 2, \ldots, n-1)$$

To interpret these and also (B) and (C) we need the generalised definition of geodesic curvature. Assume our manifold to exist in Euclidean space of sufficiently high order, and in this space through a point P of the manifold draw a vector of which the length γ is given by

$$\gamma^2 = \sum_{i=1}^{n} \gamma^2_{inn},$$

and the direction by that of the tangent to the line which passes through P and belongs to the congruence μ for which

$$\mu_r = \sum_{i=1}^{n} \gamma_{inn}\lambda_{i/r}.$$

This vector possesses the properties

(1) It is identically zero if $[n]$ is geodesic.

(2) Its projection on the plane tangent to the lines i and n is equal to the curvature of the projection of the line n on the same plane.

(3) It is normal to the line n.

For these reasons γ is called the geodesic curvature of the line n, and the congruence μ is called that of the lines of geodesic curvature of $[n]$.

Consider the lines μ of geodesic curvature of $[h]$. We have

$$\mu_{h/r} = \sum_i \gamma_{ihh}\lambda_{i/r}. \qquad (r = 1, 2, \ldots, n)$$

Multiply this by $\lambda_n^{(r)}$ and sum over r. There is obtained the equation

$$\gamma_{nhh} = \sum_r \lambda_n^{(r)} \mu_{h/r}.$$

Therefore the remaining equations (A), $\gamma_{nhh}=0$, show that the lines of geodesic curvature of every congruence orthogonal to [n] are themselves orthogonal to [n]. Similarly the equations (B) and (C) show that the cosine of the angle between a line of the congruence [h], and a line of the congruence of geodesic curvature of [n], through the same point, is b_h/ρ. Also

$$\mu_{n/r} \equiv \sum_h \gamma_{hnn}\lambda_{h/r} \equiv -\sum_h \gamma_{nhn}\lambda_{h/r}.$$

Hence, from (b), which is deducible from (B) and (C),

$$\mu_{n/r} = \rho_r/\rho.$$

Thus the congruence of lines of geodesic curvature of [n] is normal, and its orthogonal trajectories are the surfaces $\rho = \text{const.}$

In particular, if the congruence [n] is normal,

$$\gamma_{nhk} = \gamma_{nkh}, \qquad (h, k = 1, 2, \ldots, n-1)$$

and therefore all the γ's except

$$\gamma_{nhn}, \qquad (h = 1, 2, \ldots, n-1)$$

are zero.

If the system of orthogonal trajectories of [n] is $f = \text{const.}$, we have

$$f_r = \sigma\lambda_{n/r},$$

and hence

$$f_{rs} = \sigma_s\lambda_{n/r} + \sigma\sum_i \gamma_{nin}\lambda_{i/r}\lambda_{n/s}$$

$$= \sigma_s\lambda_{n/r} - \frac{\sigma\rho_r}{\rho}\lambda_{n/s},$$

and therefore since $f_{rs} = f_{sr}$ and $\lambda_{n/r} = f_r/\sigma$

we have

$$\frac{\partial}{\partial x_s}(\rho\sigma)\frac{\partial f}{\partial x_r} = \frac{\partial}{\partial x_r}(\rho\sigma)\frac{\partial f}{\partial x_s}.$$

It follows that $\rho\sigma$ is a function of f, say $F'(f)$. If we now write $F(f) = g$ we have $\rho\sigma_1 = 1$, where $g_r = \sigma_1\lambda_{n/r}$; hence finally

$$g_r = \frac{1}{\rho}\lambda_{n/r} \text{ and } g_{rs} = -\frac{1}{\rho^2}\{\rho_s\lambda_{n/r} + \rho_r\lambda_{n/s}\}.$$

Again

$$\sum_{r,s} a^{(rs)}g_{rs} = \sum_{r,s,h} \lambda_h^{(r)}\lambda_h^{(s)}g_{rs} = -\frac{1}{\rho^2}\{\sum_s \rho_s\lambda_n^{(s)} + \sum_r \rho_r\lambda_n^{(r)}\}$$

$$= -\frac{2b_n}{\rho^2} = 0,$$

and therefore, if the congruence [n] is normal, its family of orthogonal trajectories is isothermal.

Dynamical Applications.

74. The position of any material system is given with reference to a fixed frame of axes when the cartesian coordinates of every point of it are given; frequently, however, the points of the system are not independent of each other, and it may happen that the coordinates of all the points may be expressed in terms of a finite number of quantities. Thus consider a rigid straight rod in a plane. Its position, and therefore the coordinates of all its points, are known when we know the coordinates of its middle point and the angle it makes with the x axis. The position of a sphere in space is determined by the coordinates of its centre, the angle that a line fixed in it makes with the z axis, the angle that a vertical plane through the line makes with a fixed plane, and the angle that a second line, fixed in the sphere at right angles to the first, makes with the line that is in the vertical plane above mentioned and is also at right angles to the first line. Whenever a material system is thus determined in position by definite values given to certain quantities, these quantities are called the generalised coordinates of the system.

Any change in the coordinates corresponds to a change in the position of the system. Suppose the system to be in a given position. It may happen that each change in the coordinates gives a possible change in the system; on the other hand, it may happen that, owing to certain dynamical restrictions in a particular problem, some changes in the coordinates do not give possible displacements. For example, if a sphere is moving on a fixed plane with pure rolling motion, all infinitesimal displacements are excluded which do not make the displacement of the point of contact of the sphere with the plane zero (to first order of small quantities). In the first case the system is said to be *holonomic*, in the second *non-holonomic*.

75. Consider any holonomic system, and let its generalised coordinates be $x_1, x_2, \ldots, x_n$. Then, if dots denote differentiations with regard to the time, it may be proved that the kinetic energy, T, of the system is given by

$$\text{(i)} \qquad 2T = \sum_{r,s} a_{rs}\dot{x}_r\dot{x}_s,$$

where the coefficients a_{rs} are functions of $x_1, \ldots, x_n$. Again, suppose the system to be subject to external forces which depend only on position and not on velocities, then in a small displacement dx the work done by these forces is

$$\sum_r X_r dx_r,$$

where the X's are functions of $x_1, \ldots, x_n$. In this case the equations of motion of the system are

$$\text{(ii)} \qquad \frac{d}{dt}\left(\frac{\partial T}{\partial \dot{x}_r}\right) - \frac{\partial T}{\partial x_r} = X_r, \qquad (r = 1, 2, \ldots, n)$$

the equations of Lagrange*.

76. The dynamical problem involved may be regarded as completely solved (if we neglect such problems as the determination of the internal reactions of the system) when we have determined the most general values of the x's as functions of t, that satisfy the set of equations (ii), the a's and the X's being given functions of the quantities x.

Now we may regard the x's as the coordinates of a point in a general manifold of n dimensions, and then any particular solution of the equations (ii) will give the x's as functions of one parameter t. The possible values of these variables will thus determine a curve in the manifold; if this curve is given, the complete particular solution of the problem will follow as soon as one coordinate is determined as the proper function of t. Such a curve as that mentioned is called a *trajectory* of the configuration, and it is clear that we have gone far towards the complete solution when all the trajectories have been determined.

77. Any n dimensional manifold whatever may be chosen for the geometrical interpretation of the dynamical configuration. The obvious one to choose is that for which

$$ds^2 = \sum_{r,s} a_{rs} dx_r dx_s,$$

and then

$$F = \tfrac{1}{2}\dot{s}^2.$$

It is clear that, since $\sum_r X_r dx_r$ represents the work done by the forces in a small displacement, it is independent of the coordinates chosen, and hence the magnitudes X_r are a covariant system of the first order with respect to the fundamental form $\Sigma a_{rs} dx_r dx_s$. (If the

* For details of the above, the reader should see *e.g.* Whittaker, *Analytical Dynamics*, p. 32 *sqq.*

system of forces is conservative, the X's are actual derivatives of a function U). We may therefore take

$$X_r = \rho\mu_r,$$

where $\mu_1, \ldots, \mu_n$ are the coefficients of a congruence of curves. These curves are called the lines of force.

78. Expression of the equations in invariantive form.

The equations (ii) may be expressed in invariantive form in terms of an orthogonal ennuple; to express them thus we choose from the trajectories any congruence, which we take for $[n]$. Then

$$\dot{x}_r = \sigma\lambda_n^{(r)}, \text{ where } \sigma = \dot{s}_n,$$

and

$$\frac{\partial T}{\partial \dot{x}_r} = \sum_s a_{rs}\dot{x}_s = \sum_s \sigma a_{rs}\lambda_n^{(s)} = \sigma\lambda_{n/r}.$$

Hence

$$\frac{d}{dt}\left(\frac{\partial T}{\partial \dot{x}_r}\right) - \frac{\partial T}{\partial x_r} \equiv \dot{\sigma}\lambda_{n/r} + \sigma\sum_p \frac{\partial \lambda_{n/r}}{\partial x_p}\dot{x}_p - \tfrac{1}{2}\sum_{p,q}\frac{\partial a_{pq}}{\partial x_r}\dot{x}_p\dot{x}_q,$$

and, after some slight reductions, this becomes

$$\dot{\sigma}\lambda_{n/r} + \sigma^2\sum_p \lambda_n^{(p)}\lambda_{n/rp}.$$

Again

$$\lambda_{n/rp} = \sum_{ij}\gamma_{nij}\lambda_{i/r}\lambda_{j/p},$$

and therefore

$$\sum_p \lambda_n^{(p)}\lambda_{n/rp} = \sum_i \gamma_{nin}\lambda_{i/r} = -\nu_r,$$

where $[\nu]$ is the congruence of geodesic curvature of $[n]$. Hence, finally, the equations (ii) become

(iii) $$\dot{\sigma}\lambda_{n/r} - \sigma^2\nu_r = \rho\mu_r, \qquad (r = 1, 2, \ldots, n)$$

Now let ψ be any congruence whatever, multiply the typical equation (iii) by $\psi^{(r)}$ and sum over the values of r. We have

$$\dot{\sigma}\sum_r \lambda_{n/r}\psi^{(r)} - \sigma^2\sum_r \nu_r\psi^{(r)} = \rho\sum \mu_r\psi^{(r)},$$

or since $\sum_r \phi_r\psi^{(r)}$ is the cosine of the angle between ϕ and ψ,

(iv) $$\dot{\sigma}\cos(n\psi) - \sigma^2\cos(\nu\psi) = \rho\cos(\mu\psi).$$

This shows that any line perpendicular to the line n and to the line μ is also perpendicular to the line ν; in other words: *The tangent to the line of geodesic curvature of any trajectory lies in the plane determined by the tangent to that trajectory and the tangent to the line of force through the point.*

If, in particular, the forces X are all zero, $\lambda_{n/r}$ is proportional to ν_r or else $\dot{\sigma}$ and ν_r vanish identically. Now $[n]$ and $[\nu]$ are at right angles, and hence the first case gives minimal lines. Excluding these, we have the result: If the external forces are all zero, the trajectories are the geodesics of the manifold, and they are described with constant velocity.

When the field of force is not zero, we deduce, if in (iv) we take ψ to be $[h]$, any one of the congruences of the ennuple,

$$\dot{\sigma} = \rho \cos(\mu n),$$

$$\sigma^2 \gamma_{nhn} = \rho \cos(\mu h), \qquad (h = 1, 2, \ldots, n-1)$$

which are the equations in invariantive form.

79. First integrals of the system (ii).

The n equations (ii) are linear in the second derivatives of the x's with respect to t. If they are solved for these derivatives they take the form

$$\text{(v)} \qquad \ddot{x}_i = X^{(i)} - \sum_{r,s} \{rs, i\} \dot{x}_r \dot{x}_s. \qquad (i = 1, 2, \ldots, n)$$

Let $f = \text{const.}$ be a first integral of these equations, then f is a function of the quantities x and $\dot{x}$, and

$$\frac{df}{dt} = \sum_i \frac{\partial f}{\partial \dot{x}_i} \ddot{x}_i + \sum_i \frac{\partial f}{\partial x_i} \dot{x}_i$$

$$\text{(vi)} \qquad = \sum_i \frac{\partial f}{\partial \dot{x}_i} X^{(i)} + \sum_i \left\{ \frac{\partial f}{\partial x_i} \dot{x}_i - \frac{\partial f}{\partial \dot{x}_i} \sum_{r,s} \{rs, i\} \dot{x}_r \dot{x}_s \right\}.$$

This last expression must be zero identically.

Suppose in particular that f is a polynomial in the first derivatives $\dot{x}$, and let the terms of highest degree in these derivatives be u. Then, since (vi) vanishes identically, the coefficients of the various quantities $\dot{x}_r$ in it must vanish separately, and hence

$$\text{(vii)} \qquad \sum_i \left\{ \frac{\partial u}{\partial x_i} \dot{x}_i - \frac{\partial u}{\partial \dot{x}_i} \sum_{r,s} \{rs, i\} \dot{x}_r \dot{x}_s \right\} \equiv 0.$$

Therefore $u = \text{const.}$ must be an integral of the equations (ii) when all the X's are made zero. That is to say, to any polynomial first integral of the general set (ii) there corresponds a homogeneous polynomial first integral of the differential equations for geodesics, in the manifold for which

$$ds^2 = \sum_{r,s} a_{rs} dx_r dx_s.$$

Assume that

$$u \equiv \sum_{r_1, \ldots, r_m} c_{r_1 \ldots r_m} \dot{x}_{r_1} \ldots \dot{x}_{r_m},$$

then the system c is covariant of the rth order, and (vii) becomes

$$\sum_{r_1, \ldots, r_{m+1}} c_{r_1 \ldots r_m r_{m+1}} \dot{x}_{r_1} \ldots \dot{x}_{r_{m+1}} \equiv 0.$$

80. Homogeneous linear first integrals of the equations for geodesics.

For example, suppose that m is unity, then

$$u = \sum_r c_r \dot{x}_r,$$

and (vii) becomes

$$\sum_{r,s} c_{rs} \dot{x}_r \dot{x}_s \equiv 0.$$

Therefore
$$c_{rs} + c_{sr} = 0. \qquad (r, s = 1, 2, \ldots, n)$$

These conditions show that the quadratic form must admit the infinitesimal transformation

$$\sum_r c^{(r)} \frac{\partial f}{\partial x_r},$$

and hence the necessary and sufficient condition that the equations for geodesics of a manifold have a linear first integral is that the quadratic form of the manifold admit an infinitesimal transformation.

If the general system (ii), where the X's are not zero, also has the linear solution $u = \text{const.}$, the remaining terms from (vi) give

$$\sum_r c_r X^{(r)} = 0,$$

and hence the additional condition is that the path curves for the infinitesimal transformation must be orthogonal to the lines of force.

81. Quadratic first integrals.

Systems (ii) that possess a quadratic first integral are of particular interest, for a reason that will appear later. Suppose the first integral to be

$$\sum_{r,s} c_{rs} \dot{x}_r \dot{x}_s = \text{const.}$$

Then c is a covariant system of the second order with reference to $\sum_{r,s} a_{rs} dx_r dx_s$ for which $c_{rs} = c_{sr}$, and our conditions are

$$\sum_{r,s,t} c_{rst} \dot{x}_r \dot{x}_s \dot{x}_t \equiv 0, \qquad \sum_{r,s} c_{rs} X^{(r)} \dot{x}_s \equiv 0.$$

These give the equations

$$\text{(viii)} \qquad c_{rst} + c_{str} + c_{trs} = 0, \qquad (r, s, t = 1, 2, \ldots, n)$$

$$\text{(ix)} \qquad \sum_r c_{rs} X^{(r)} = 0. \qquad (s = 1, 2, \ldots, n)$$

Hence the necessary and sufficient conditions that the equations for geodesics possess a homogeneous quadratic integral are the equations (viii). If, in addition, a general dynamical system for which the X's are not zero possess that integral, the equations (ix) must also be satisfied for non-zero values of the X's, and hence the discriminant of the quadratic $\sum_{r,s} c_{rs}\dot{x}_r\dot{x}_s$ must vanish. If this discriminant vanish, the ratios of the X's can be determined, in general uniquely, from the equations (ix).

If the quadratic integral is not homogeneous, let it be

$$\sum_{r,s} c_{rs}\dot{x}_r\dot{x}_s + \sum_r b_r\dot{x}_r + u = \text{const.}$$

In this case the system must possess the linear integral $\sum_r b_r\dot{x}_r = \text{const.}$, and hence also the quadratic integral $\sum_{r,s} c_{rs}\dot{x}_r\dot{x}_s + u = \text{const.}$ The conditions for this last are (viii) and

$$\text{(x)} \qquad \sum_r c_{rs}X^{(r)} + u_s = 0. \qquad (s = 1, 2, \ldots, n)$$

One solution of (viii) is obvious, for we know that the first derived system of the coefficients a_{rs} of the quadratic form is identically zero. Hence one quadratic first integral of geodesics is $\sum_{r,s} a_{rs}\dot{x}_r\dot{x}_s$, and since $\sum_r a_{rs}X^{(r)} \equiv X_r$, the equations (x) give $X_r + u_r = 0$, that is to say, the forces X must be actual first derivatives of a function $-u$. The system of forces is therefore conservative, and the first integral is the energy integral.

In addition to the solution $c_{rs} = a_{rs}$ the general form does not possess a quadratic first integral for geodesics, and all the forms which do possess such integrals have not yet been determined, though many classes of such forms have been obtained. The case of $n = 2$ has been completely solved by Liouville.

82. Systems which have the same trajectories.

Suppose that we have any dynamical system for which the equations are the set (ii). Any solution whatever is given by definite expressions for the n variables $x_1, \ldots, x_n$ as functions of the single parameter t. Thus to each solution there corresponds a curve in any n dimensional space, the trajectory of the solution. Now suppose that we have any other dynamical system

$$\text{(xi)} \qquad \frac{d}{dt_1}\left(\frac{\partial U}{\partial x_r'}\right) - \frac{\partial U}{\partial x_r} = Y_r, \qquad (r = 1, 2, \ldots, n)$$

where $2U = \sum_{r,s} c_{rs}x_r'x_s'$, and accents denote derivatives with regard to t_1. We enquire what are the conditions that the trajectories of this system are the same as those of the former system. If the trajectories are the same in the two systems, let $x_i = f_i(t)$, $(i = 1, 2, \ldots, n)$, be a solution of the former; then if $x_i = F_i(t_1)$ is the corresponding solution of the latter, that is to say the solution that gives the same trajectory, these two expressions for x_i must become the same when we choose for t an appropriate function of t_1. We write $t = \theta(t_1)$ and take $\sum_{r,s} a_{rs} dx_r dx_s$ as the fundamental form, and we also make the assumption that the discriminants of the two quadratic forms do not vanish. Confining ourselves to a particular trajectory, we have

$$x_r' = \dot{x}_r \theta', \quad x_r'' = \dot{x}_r \theta'' + \ddot{x}_r \theta'^2,$$

also

$$\ddot{x}_r + \sum_{p,q} \{pq, r\} \dot{x}_p \dot{x}_q = X^{(r)},$$

and (xi) is

$$\frac{d}{dt_1}\left(\sum_s c_{rs} x_s'\right) - \tfrac{1}{2} \sum_{p,q} \frac{\partial c_{pq}}{\partial x_r} x_p' x_q' = Y_r,$$

or

$$\sum_s c_{rs} x_s'' + \sum_{s,p} \frac{\partial c_{rs}}{\partial x_p} x_s' x_p' - \tfrac{1}{2} \sum_{p,q} \frac{\partial c_{pq}}{\partial x_r} x_p' x_q' = Y_r.$$

If we substitute in this for x', x'', $\ddot{x}$, and replace the quantities $\frac{\partial c_{rs}}{\partial x_p}$ by the covariant derivatives with respect to the fundamental form $\sum a_{rs} dx_r dx_s$, this equation (xi) becomes, after a few reductions,

$$\text{(xii)} \quad \sum_s c_{rs} \dot{x}_s \theta'' + \sum_s c_{rs} X^{(s)} \theta'^2 + \tfrac{1}{2} \theta'^2 \sum_{p,q} (c_{rpq} + c_{rqp} - c_{pqr}) \dot{x}_p \dot{x}_q = Y_r.$$

Now we have n equations of type (xii), and if any three of them were independent we could eliminate θ'', θ'^2, and thus remain with an equation involving only the first derivatives $\dot{x}$, which would be the same for every trajectory; this is impossible, and it follows that two at most of the equations (xii) can be independent. One solution of these equations is immediately obvious. If θ'' is zero, it readily follows that c_{rs} is Ca_{rs} where C is a pure constant, and then the Y's bear a constant ratio to the X's. We neglect this case, and consider that in which all the Y's are zero. There can be now only one equation (xii), and therefore

$$\text{(xiii)} \quad \frac{\sum_s c_{rs} \dot{x}_s}{\sum_s c_{ks} \dot{x}_s} = \frac{\sum_s c_{rs} X^{(s)} + \tfrac{1}{2} \sum_{p,q} (c_{rpq} + c_{rqp} - c_{pqr}) \dot{x}_p \dot{x}_q}{\sum_s c_{ks} X^{(s)} + \tfrac{1}{2} \sum_{p,q} (c_{kpq} + c_{kqp} - c_{pqk}) \dot{x}_p \dot{x}_q}$$

must be an identity, for all values of r and k. If we multiply up, and equate coefficients of the derivatives $\dot{x}$ on both sides, we have in the first place

$$\left(\sum_s c_{rs} \dot{x}_s\right)\left(\sum_s c_{ks} X^{(s)}\right) \equiv \left(\sum_s c_{ks} \dot{x}_s\right)\left(\sum_s c_{rs} X^{(s)}\right).$$

Hence, since the discriminant of U does not vanish, we must have $\sum_s c_{ks} X^{(s)} = 0$ for all values of k, and therefore finally $X^{(s)} = 0$.

83. Representation of one manifold on another with correspondence of geodesics.

Therefore, if two dynamical systems have the same trajectories, and if these trajectories are geodesics for one system, they are geodesics also for the other, and we now consider the question of representing one manifold on another so that the geodesics of the one correspond to geodesics of the other. The equation (xiii) gives

$$(\sum_s c_{rs}\dot{x}_s)\left[\sum_{p,q}(c_{kpq}+c_{kqp}-c_{pqk})\,\dot{x}_p\dot{x}_q\right]\equiv(\sum_s c_{ks}\dot{x}_s)\left[\sum_{p,q}(c_{rpq}+c_{rqp}-c_{pqr})\,\dot{x}_p\dot{x}_q\right].$$

This equation shows that $\sum_s c_{rs}\dot{x}_s$ must be a factor of

$$\sum_{p,q}(c_{rpq}+c_{rqp}-c_{pqr})\,\dot{x}_p\dot{x}_q,$$

and the remaining factor must be symmetrical. Hence we have

$$\sum_{p,q}(c_{rpq}+c_{rqp}-c_{pqr})\,\dot{x}_p\dot{x}_q\equiv(\sum_s c_{rs}\dot{x}_s)\,(\sum_p b_p\dot{x}_p)$$

for all values of r, and the b's are quantities to be determined.

From this equation, by equating coefficients of $\dot{x}_p\dot{x}_q$ on both sides, and noting that $c_{pq}=c_{qp}$ we have

$$\text{(xiv)}\qquad 2\,(c_{rpq}+c_{rqp}-c_{pqr})=c_{rp}b_q+c_{rq}b_p,$$

for all values of p, q, r.

Let a, c, be the discriminants of the two forms $\sum a_{rs}dx_rdx_s$, $\sum c_{rs}dx_rdx_s$, then c/a is an invariant. Also, if the cofactor of c_{rs} in c is $ak^{(rs)}$, then $nc/a=\sum_{r,s}c_{rs}k^{(rs)}$, and therefore the system $k^{(rs)}$ is contravariant. Multiply (xiv) by $ak^{(rq)}$ and sum over the values of r; the result is

$$2a\sum_r(c_{rpq}+c_{rqp}-c_{pqr})\,k^{(qr)}=cb_p,$$

for all values of q except $q=p$. If $p=q$ the right side becomes $2cb_p$. When these equations are summed over all values of q there is obtained the result

$$(n+1)\,cb_p=2a\sum_{q,r}(c_{rpq}+c_{rqp}-c_{pqr})\,k^{(qr)},$$

and since $k^{(qr)}=k^{(rq)}$ this becomes

$$(n+1)\,cb_p=2a\sum_{q,r}k^{(qr)}c_{qrp}.$$

Now
$$\frac{c}{a}=\frac{1}{n\,!}\sum_{r,s}\epsilon^{(r_1\ldots r_n)}\epsilon^{(s_1\ldots s_n)}c_{r_1s_1}c_{r_2s_2}\ldots c_{r_ns_n}.$$

and therefore

$$(c/a)_p = \frac{1}{n\,!}\left\{\sum_{r,\,s} \epsilon^{(r_1\ldots r_n)}\epsilon^{(s_1\ldots s_n)} c_{r_1 s_1 p} c_{r_2 s_2} \ldots c_{r_n s_n} + \overline{n-1} \text{ similar terms}\right\}$$

$$= \frac{1}{n\,!}\left\{\sum_{r_1,\,s_1} (n-1)\,!\; k^{(r_1 s_1)} c_{r_1 s_1 p} + \overline{n-1} \text{ similar terms}\right\}$$

$$= \sum_{r,\,s} k^{(rs)} c_{rsp}.$$

Hence $$(n+1)\,cb_p = 2a\,(c/a)_p,$$

and therefore $$b_p = \frac{2}{n+1}\frac{\partial}{\partial x_p}\left(\log\frac{c}{a}\right).$$

Thus the b's are the derivatives of the function $\frac{2}{n+1}\log(c/a)$, and from (xii) we have

$$\theta'' + \tfrac{1}{2}\theta'^2 \sum_p b_p \dot{x}_p = 0,$$

and therefore $$\theta' = C\,(a/c)^{\frac{1}{n+1}} = \mu, \text{ say,}$$

where C is an arbitrary constant.

If we write $A_{rs} \equiv \mu^2 c_{rs}$ we deduce at once

$$A_{rst} + A_{str} + A_{trs} = 0,$$

and therefore the set (ii) admit the quadratic first integral

$$\sum_{r,\,s} A_{rs}\dot{x}_r\dot{x}_s = \text{const.}$$

84. Again, we write

$$\sum_s (c_{rs} - \rho a_{rs})\,\lambda^{(s)} = 0. \qquad (r = 1,\, 2,\, \ldots,\, n)$$

If this set of equations gives a system of non-zero values for the λ's, the determinant $\| c_{rs} - \rho a_{rs} \|$ must be zero. To each root of this equation of the nth degree in ρ there corresponds a set of values for the λ's, and, if the roots are distinct, these n sets determine an orthogonal ennuple. If the roots are not distinct, an orthogonal ennuple may still be determined in this way, the only difference being that k congruences appertain to a root of multiplicity k of the equation $\| c_{rs} - \rho a_{rs} \| = 0$.

If we express our quantities in terms of an ennuple thus determined, we have

$$c_{rs} = \sum_h \rho_h \lambda_{h/r} \lambda_{h/s},$$

and when these values are substituted in the equations of condition (xiv), we get a set of equations which reduces finally to the following set:

$$(\alpha)\qquad (\rho_h - \rho_i)\,\gamma_{hij} = 0, \qquad (h \neq i \neq j)$$

$$(\beta)\qquad 2\,(\rho_i - \rho_j)\,\gamma_{iji} = \frac{\partial \rho_i}{\partial s_j}, \qquad (i \neq j)$$

$$(\gamma)\qquad \frac{\partial\,(\mu\rho_i)}{\partial s_j} = 0, \qquad (i \neq j)$$

$$(\delta)\qquad \frac{\partial\,(\mu\rho_i)}{\partial s_i} + \rho_i \frac{\partial \mu}{\partial s_i} = 0,$$

$$(h,\ i,\ j = 1,\ 2,\ \ldots,\ n).$$

Suppose that the roots ρ are all different, then, from (α), all the γ's with three distinct suffixes are zero, and therefore the n congruences are normal. We take the normal surfaces of these congruences as the coordinates x, and then the fundamental quadratic form becomes

$$\sum_i H_i^2 dx_i^2,$$

and $ds_i = H_i dx_i$. The set of equations (α) are now satisfied, and the remainder are

$$(\beta_1)\qquad 2\,(\rho_i - \rho_j)\frac{\partial}{\partial x_j}(\log H_i) + \frac{\partial \rho_i}{\partial x_j} = 0, \qquad (i \neq j)$$

$$(\gamma_1)\qquad \frac{\partial}{\partial x_j}(\mu\rho_i) = 0, \qquad (i \neq j)$$

$$(\delta_1)\qquad \frac{\partial}{\partial x_i}(\mu\rho_i) + \rho_i \frac{\partial \mu}{\partial x_i} = 0.$$

These equations are integrated without difficulty. From (γ_1) we have $\mu\rho_i = \phi_i$, where ϕ_i is a function of x_i alone, and then, from (δ_1), $\mu\phi_i$ is independent of x_i. Hence if $F \equiv \phi_1\phi_2 \ldots \phi_n$, μF is independent of x_i, and therefore, since it is symmetrical, it is a pure constant, C. Hence $\mu = C/F$, and the equation (β_1) becomes

$$2\,(\phi_i - \phi_j)\frac{\partial}{\partial x_j}(\log H_i) + \frac{\phi_i}{\phi_j}\frac{\partial \phi_j}{\partial x_j} = 0,$$

and therefore on integrating we see that $\phi_j H_i^2/(\phi_i - \phi_j)$ is independent of x_j. Let P_i denote the product $\prod_{j=1}^{n}{}' (\phi_i - \phi_j)$, where the value $j = i$ is excluded; then FH_i^2/P_i is a function of x_i alone, and by properly choosing a function of x_i as a new variable instead of x_i we may make its value C/ϕ_i.

The quadratic form $\Sigma a_{rs} dx_r dx_s$ can therefore be transformed so as to be

$$\sum_i \frac{CP_i}{F\phi_i} dx_i^2.$$

Also $$c_{rs} = \sum_h \rho_h \lambda_{h/r} \lambda_{h/s} = 0, \qquad (r \neq s)$$

and $$c_{ii} = \rho_i H_i^2 = P_i \,;$$

therefore the corresponding form is

$$\sum_i P_i dx_i^2.$$

We thus have the complete solution of the problem of representation with correspondence of geodesics in the case when the ρ's are all distinct. The case when some of the ρ's are alike involves modifications but can also be fully solved.

The general problem with which we started, in which the forces are not zero, has not yet been fully solved. The method of the Absolute Calculus removes all analytical difficulties not inherent in the problem, and it is possible that that method will ultimately give the general solution*.

* On this problem see: Painlevé, *Liouville's Journal*, Ser. v. Vol. x. (1894); Levi-Civita, *Annali di Math.* Ser. II. Vol. XXIV. (1896).

Printed in the United States
By Bookmasters